Study Guide to accompany

HUMAN DEVELOPMENT

Fifth Edition

James W. Vander Zanden

Prepared by
Thomas L. Crandell, Ph.D.
Corinne H. Crandell, M.S.

McGraw-Hill Inc

Study Guide
HUMAN DEVELOPMENT

234567890 MAL MAL 909876543

ISBN 0-07-067000-5

The editors were Nomi Sofer and James Belser;
the production supervisor was Richard A. Ausburn.
Malloy Lithographing, Inc., was printer and binder.

Contents

Study chapts 5, 9, 13, 15, 19

5/16/94 Test

17
18
19 - Later adulthood

243

iii

Preface

To the Student:

Human Development is a course in psychology which will have practical and lasting effects on your personal journey through life. You will discover new and exciting insights into your own personal development as well as that of your family, friends, and acquaintances. The fifth edition of *Human Development* by James W. Vander Zanden has been revised to incorporate the latest research findings and theories; and in the course of revision, the publisher has taken extra care to ensure that the book is readable and easy to use.

Nevertheless, like any other academic course in college, the material in this textbook must be learned and submitted to memory for future recall on tests. We have written this *Study Guide* to help you accomplish this task and to achieve the best grades possible.

We also recognize that students differ with respect to their preferred style of learning. So we encourage you to continue using the study method which you have found most successful. At the same time, having access to this *Study Guide*, which we designed specifically for the 5th edition of the Vander Zanden text, should help you to study more effectively and increase your confidence in class and achieve even higher grades.

The conceptual scheme we used for preparing this *Study Guide* is based on sound principles of human learning; thus all students, regardless of age, experience, or ability, should profit from using the guide. This would be especially true for any learners who have been away from the formal classroom for a period of time and could use assistance with sharpening their study skills and habits.

Welcome to the field of human development. Enjoy your course!

How to Use this *Study Guide*

The student *Study Guide* is organized to coincide with the chapters of your textbook. Each chapter contains the following sections.

Introduction

The first section--the introduction--serves as a brief orientation to the corresponding material in the chapter. You can use the introduction in two ways: (1) to help you organize your thinking before reading the chapter or (2) to review what you have learned after having read the chapter.

Chapter Outline

The chapter outline gives all the headings and subheadings within the text chapter and thus shows you the complete structure, or framework, of the chapter material. Enough space is left between entries for you to make brief notes.

The chapter outlines will also help you preview each chapter and will clarify the relationship among topics and subtopics within the chapter.

Learning Objectives

The learning objectives provide you with a listing of the major topics for each chapter. Each objective corresponds to a major principle or concept discussed in the main text. By carefully reading the learning objectives *before* you read the text chapter, you will know in advance what is expected of you. We have left space between objectives so that you can take brief notes as you read the textbook chapter. Use these notes as a foundation for your classroom lecture notes and for organizing your thoughts.

Furthermore, the objectives can be particularly beneficial if your instructor gives essay-type questions or examinations, since essay questions may often resemble these objectives. Writing out answers to some of the objectives will give you practice in dealing with essay exams.

Key Terms

This section of key terms provides you with a list of the important vocabulary that you need to learn in order to master the subject matter in the textbook. The key terms are listed in alphabetical order, and space is provided for you to make short notes of definitions or examples. For each key term, a page number in the text is cited; this is where the definition or explanation is located. Once you have made notes of definitions, examples, or both, the key terms sections will be an excellent reference to review for examinations.

Note that when key terms are used in the learning objectives (the previous section of each *Study Guide* chapter), they are set in *italics* at their first appearance.

Self-Tests

One of the most effective ways to prepare for in-class examinations is to test yourself using questions similar to those you might find on an actual exam. This is the purpose of the self-test section of the *Study Guide*. Each self-test contains a carefully prepared set of sample questions--multiple-choice, matching, and completion items--that test your understanding of facts and concepts in the textbook chapter. You should test yourself on these questions; that is, try to answer them from memory *after* you have studied the material in the textbook chapter.

Answer Key

After you finish the self-test, use the answer key at the end of each chapter in the *Study Guide* to check your answers. The results of the self-test should then be used to determine whether you need to spend more time reading and reviewing the chapter. To help you restudy the questions you find troublesome, we have provided a text page reference for each answer.

A Sample Guide for
Studying, Improving Your Memory, and Taking and Passing Tests

---

A Note on Learning Styles

Keep in mind that not everyone learns in the same way. For example, some students understand information better when they see it in written form--they learn best from the textbook, the *Study Guide*, lecture notes and handouts, etc. Other students prefer to hear information--they learn best from listening to lectures, making and listening to tape recordings of lectures, making and listening to recordings of their own notes from learning objectives and key terms sections of the *Study Guide*.

Some students may concentrate better when studying while listening to music; others may find the music distracting or even annoying. Of course, some people learn best when studying with other students in small groups where answers can be discussed. Many people prefer to study alone.
As we said earlier, if you are using a study method that works for you and you are satisfied with the results you are getting (i.e., good grades), then stick with it. However, if you are struggling and seem to be spending too much time on your homework without satisfactory results, then you need to look at some alternative methods. For example, if you know information is easier for you to learn when you *hear* words rather than when you *see* them, you should be aware that most textbooks have been recorded on tape. Check with your college's library to see if *Human Development* is available on tape format.

If you know that you have a visual learning disability that can be substantiated by a professional, you are eligible to get recorded books free of charge from the Library for the Blind in Washington, D.C. You may need to contact a counselor on your campus who assists students with learning disabilities to make this request. You may also be able to work directly with a tutor on your campus (this service is often provided free to you). Some colleges provide note-takers for students who need this extra assistance.

You know yourself better than anyone else does; don't think that because one study technique works for a particular student it will necessarily work for you. By following the learning/study method that is most helpful to you, you will make the material more memorable and more retrievable at test time.

Regardless of your preferred method or *style* of learning, there are some general guidelines for learning and improving your memory, which we present for your consideration.

Using this *Study Guide* Most Effectively

From our years of teaching Human Development to different types of learners, we have found that, in general, a study guide has proven most effective when used in the following sequence of steps.

Step 1: Preview. Before reading each chapter in the textbook, read the introduction to the chapter in the *Study Guide*, and examine the key terms.

Next, read and familiarize yourself with the chapter outline and the learning objectives in the *Study Guide*. This will help you see the scope and direction of the material in the text. The *Study Guide* serves as an advance organizer (a bridge between old and new learning), enabling you to anticipate important issues, facts, and concepts in the textbook.

This preview should take only a few minutes.

Step 2: Read. Break down the textbook into small study units, and read it one unit at a time. You can use the chapter outlines in the *Study Guide* to establish your "chunks" or study units. Preferably, each unit, and thus each learning session, should cover only about 7 to 9 pages of the text. As you read, use the blank space provided in the *Study Guide* to write brief definitions of the key terms and brief notes on each of the learning objectives.

Merely reading a chapter passively does not mean that you have mastered it. Everything may seem to make sense as you read it; however, if you try to recall the material or summarize the main points, you will frequently find that terms, concepts, and names are not retrievable. On the other hand, by defining key terms and responding to the learning objectives, you will make the text more significant and hence easier to recall on a test (or, for that matter, in appropriate situations in real life).

Step 3: Review and Test Yourself. Review what you have learned; then take the self-tests in the *Study Guide*. Check your answers against the answer key and correct your mistakes. Take the self-tests seriously. If you could not answer a question, do not merely look up the answer in the key. For each answer in the key, a text page reference is given--use this reference to find the textbook passage that answers the question, and reread that passage. Then reanswer the question. In this way, you will shore up any areas that need additional review and help store the information in long-term memory.

Step 4: Reread. Reread the textbook chapter, and then go through the self-test again. As you go through Step 4, you should find that the chapter material is more "connected"--and therefore much easier to remember.

Improving Your Memory

Try to *motivate* yourself before you begin to study. With your favorite subject, this is no problem. With a subject you don't like, try to set specific goals for your study period.

Select the material which you feel is the most important to be studied. What has your teacher stressed in class? Check your notes and handouts. What is most likely to be on the test?

Organize the facts and concepts. You can't learn disorganized information. Organization is like a key which unlocks the information in your mind. Use categories or association. Are you learning the steps in a process? A list of causes?

Recite. Go over the material several times until you have mastered it. You have to do this to completely remember any material. Use your own words in studying: *paraphrasing* definitions or writing brief *summaries* of material is an excellent study method.

Review often. Several short periods with your study material are much more effective than one prolonged session.

Taking and Passing Classroom Tests

Reducing Test Anxiety

Test anxiety can be defined as a feeling of helplessness before or during a test.

It is important to realize that examinations cause many people to become anxious and that this is normal. Complete freedom from test anxiety is unattainable; and even if it were attainable, it would probably not be desirable. Low to moderate test anxiety before a test actually tends to have a positive effect on test performance.

However, if your test anxiety is so high that it prevents you from demonstrating what you have learned, then it becomes a problem and should be addressed. Unfortunately, such anxiety is usually a complex problem, which often cannot be traced to any single cause. Personality traits (such as a tendency to take risks), emotional states (such as a negative outlook and fear of failure), and personal needs and priorities (such as overemphasis on grades) can sabotage your performance on tests.

If you consistently experience test anxiety, we recommend that you make an appointment with someone in the college or university counseling center to discuss the problem and work out a procedure for dealing with it. College counselors are trained to help you assess the cause or causes of your test anxiety and to provide you with strategies for reducing it and for improving your performance on tests.

Learning to Be the Best Test Taker You Can Be

Students who are good test takers are able to prepare for tests efficiently and to take advantage of the characteristics of tests.

For example, different tests (such as multiple-choice tests and essay tests) have different properties, which students can be taught to recognize. In fact, your college library or bookstore should be able to provide materials dealing with the nature of tests and "test smarts"; and we can recommend one such reference: a practical, pocket-sized book by Jason Millman entitled *How to Take Tests* (Cornell Publishing, Cornell University, Ithaca, New York). Some colleges also offer workshops or study sessions on how to take different kinds of tests. Remember that instructors want your test scores to reflect what you have learned. If you do poorly because you do not understand the nature of the test, then the test becomes an obstacle to accurate assessment rather than a tool for assessment. Familiarizing yourself with various kinds of tests will help ensure that your grade will be determined by your learning.

Following are some strategies we suggest to help you prepare for and take tests, so that an examination will become a genuine opportunity for you to demonstrate what you have learned.

Intellectual Preparation

Preparing for a test has intellectual, emotional, and physical aspects. Let's look at some strategies for *intellectual* preparation.

- Attend classes.

- Follow our step-by-step procedure for using *Human Development* and the *Study Guide*. Remember that this procedure can be adapted for use with other textbooks and supplements.

- Schedule regular study sessions in a specific, quiet place; and set small, reachable goals for each study session.

- Study relevant quizzes and tests you have already taken.

- Become familiar, in advance, with the purpose and format of the test. Ask your instructor to give you information on what types of questions will appear.

- See yourself--realistically--as succeeding on the test.

Emotional and Physical Preparation

Now, let's look at some strategies for preparing yourself emotionally and physically.

- Appreciate the usefulness of the test.

- Relax.

- Concentrate.

- Get a good night's sleep before the test.

- Eat a good meal before the test.

- Avoid sugars before the test, and consume *no* caffeine.

- Arrive early for the test, and come prepared with all necessary supplies (such as #2 pencil, pens, paper).

Taking the Test: General Strategies

Use time wisely. Since the time allowable for taking the test is limited, it's important to use that time efficiently.

- Find out how long you have to complete the test.

- Look over the entire test briefly before you start to answer any questions. Find out which items yield the most points.

- Begin to work as rapidly as is possible with some reasonable assurance of accuracy.

- At the outset, omit items that stump you, or just take a guess. If you have enough time when you've completed the rest of the test, remember to return to these items.

- Work immediately, and quickly, on the items which will yield the most points.

- If you become too nervous to work, stop briefly and use some relaxation techniques to calm yourself.

- If you have any time left when you've finished the test (including any difficult items you skipped at the beginning), use it to reconsider and improve your answers. As a rule, however, don't change an answer unless you are absolutely certain that you misread the question initially or that you missed some important aspect of it. Your first answer is often the correct one.

Read all directions and questions carefully. Students often lose points simply because they haven't followed directions or because they have misinterpreted questions.

- Before you start writing, become familiar with the test directions.

- Pay particular attention to the directions that most influence how you will take the test.

- Ask the examiner for clarification when necessary.

- Keep the test directions in mind while you are answering the questions.

- Be careful to read each question "as is"--not as you might like it to be.

- Pay attention to any vocabulary terms that appear in the questions. If you are allowed to make your own notes on your exam paper, sometimes it helps to circle or highlight vocabulary terms in a question; this can help you pinpoint what the question is asking for.

- If you can write your own notes on the exam paper, it may be helpful in multiple-choice to cross out answers that you have eliminated as incorrect, and in matching sections to cross off items you've already paired up.

Taking the Test:
Strategies for Specific Types of Questions

Multiple-choice items. The typical multiple-choice format consists of an incomplete sentence with several options for completing it or a question with several possible answers. (You may have to circle the correct choice or write its identifying letter or number in an answer space.)

- Read the fragment or question carefully, anticipate the answer, and then look for your anticipated answer among the choices.

- If the choices do not include the answer you anticipated, consider all the alternatives using a process of elimination. It can be helpful to treat the item as a "completion" question, covering all the choices and then uncovering one at a time.

- If, in a four-choice format, you have eliminated two of the choices but are undecided about the remaining two, treat each of the remaining two as a "true-false" question.

- Relate each option to the question.

Matching items. The typical matching format consists of two sets of items to be paired off. They might be in side-by-side columns; or one set might be inside a box, as in this *Study Guide*; or there may be some other setup. (Arrangements for indicating the answers vary. In this *Study Guide*, for instance, each item in one of the sets is followed by an answer space.)

- As always, read the directions carefully.

- Count the number of items in each set.

- Determine the relationship between the two sets.

- Try the first item. If you can't find its "partner" in the second set, skip to the next item. Keep skipping until you find one matching pair, then go on until you find another pair, and so on.

- When you have matched all the items you know, use a process of elimination for the remaining items. If you are allowed to write your own notes on the exam paper, cross out the items you have already matched.

Completion items. The typical completion, or fill-in item, is a sentence with one or more blanks; you are to make the sentence read correctly by supplying whatever is needed in each blank. (You may be asked to write your answer in the blank itself or in a separate answer space.)

- Give a general answer if you don't know the specific answer.

- Examine the sentence for grammatical clues. (For example, *a* or *an* preceding a blank tells you that the answer is singular; *these* or *those* tells you that the answer is plural.)

True-false items. The typical true-or-false item is a statement which you are to identify as correct or incorrect. (You may have to write *T* or *true*, or *F* or *false*, in an answer space; or check off or circle *T* or *F*.)

- Remember the odds (50-50).

- As always, read each item carefully.

- Look for qualifiers (*not, new, recent,* etc.)

- Watch for absolute terms (*always, never, all, none, every,* etc.). Items using absolute terms are usually false.

- Watch for conditional terms (*some, few, occasionally, sometimes*). Items using conditional terms are usually true.

Essay items. An essay item may be phrased as a question ("Why did Freud believe that . . .?") or as an imperative ("Explain why Freud believed that . . ."). You are to write a full answer. Often, your answer will be graded not only for content but also for correct grammar and for the logic of your presentation. (Space may be provided for each answer; or all the essay items may be on a printed sheet, with answers to be written in an examination booklet or on your own paper.)

- When a test has more than one essay item, read each one carefully.

- If you are allowed to make notes on the exam paper, jot down beside each essay item the relevant points that occur to you.

- If you can make notes on the exam paper, you may also want to highlight or circle parts of the question that indicate exactly what you are being asked to write about.

- Analyze the verbs in the item: *contrast, compare, describe, list, explain,* etc. Circle them for emphasis if that is permitted.

- Organize your answer before you start writing.

- If you are not sure of the best answer, quickly write down all your ideas.

- If you do not have enough time to write a full essay answer, give your answer in outline form.

- When you have finished an essay item, read it over. Check to be sure that you have followed each of the direction verbs you identified (you may have circled these, as noted above).

- WRITE LEGIBLY.

"Bubble sheets" and *optical scanner sheets.* These are not, of course, types of questions; nevertheless, they are formats that you should know how to deal with. Two hints:

- Before turning in your exam, make sure that you have filled in all the spaces.

- If you skip any items on the test, be sure to complete the remaining items in the correct order.

A Final Comment

While this *Study Guide* is designed to help you learn the information in the main textbook, remember there are no shortcuts to effective learning and good grades--motivation, effort, practice, and endurance are the keys that will unlock your potential.

Use this *Study Guide* as a learning tool to help you achieve your goal. Therefore, mark up the pages; make notes on the chapter outlines; make notes of definitions and examples of key terms; jot down your thoughts about the learning objectives; and write in your answers to the self-tests.

As previous students and teachers of many college courses, we have found that our suggestions do help, and we share them with you along with our welcome to the Human Development course and our best wishes for your success.

Acknowledgments

To all of the special people who have helped us prepare this *Study Guide*, we thank you--especially:

Jim Crandell and Colleen Crandell, two recent college graduates who assisted and gave us a "student" perspective. Our editors at McGraw-Hill: Naomi Sofer, for steering us through the white waters of textbook publishing and always being there when we had questions. Jane Vaicunas, who continues to support us in our endeavor to help students learn about psychology by providing them with *usable* instructional materials.

Thomas L. Crandell, Ph.D.
Corinne H. Crandell, M.S.

Chapter 1

Introduction

INTRODUCTION

Chapter 1 offers an overview of the developmental perspective, outlines basic concepts, and identifies various research methods. Several important topics are presented, including the following:

- The four primary concerns of developmental psychologists: describing, explaining, predicting, and controlling development

- A structure for organizing information about human development from various standpoints, which allows one to study this diverse field effectively

- Cultural and historical influences on human development and how these forces shape our conception of the life cycle

- A detailed description of the various research methods used by scientists to study human nature, as well as the framework of ethical considerations for conducting research on human subjects

CHAPTER OUTLINE

I. The Major Concerns of Science

II. A Framework for Studying Development

    A. Major Domains of Development

    B. Processes of Development

    C. The Context of Development

    D. The Timing of Developmental Events

III. Partitioning the Life Span: Cultural and Historical Perspectives

    A. Cultural Variability

    B. Historical Conceptions of the Life Span

IV. The Nature of Developmental Research

    A. The Longitudinal Method

    B. The Cross-Sectional Method

    C. The Case-Study Method

    D. The Experimental Method

    E. The Social Survey Method

    F. The Naturalistic Observation Method

    G. The Cross-Cultural Method

V. Ethical Standards for Human Development Research

LEARNING OBJECTIVES

After completing Chapter 1, you should be able to:

1. Explain what is meant by the study of *human development*. (page 4)

    Order
    Change
    Passage

2. Describe the four major issues focused on by scientists who study developmental psychology: (5)

    a. *Description*

    b. *Explanation*

c. *Prediction*

d. *Control*

3. Name the three fundamental domains within which *development* takes place, and give an example of each. (6)
   a. *Nature*   *Physical development*
   b. *Nurture*   *Cognitive development*
   c. *Interaction - Heredity & environment intertwine*
      Psychosocial development

4. Define the processes of *growth, maturation,* and *learning*; and provide an example of each as it relates to our understanding of human development. (6-7)
   a.

   b.

   c.

5. Outline the *ecological approach* toward understanding human development. (8)

6. Describe the four levels of environmental influence on human development as presented in the *ecological approach*, and give an example of each. (8-9)
   a.

   b.

   c.

   d.

7. Describe the following influences on human development: (10-11)
   a. normative age-graded influences

   b. normative history-graded influences

   c. nonnormative life events

8. Explain how age operates as a reference point in helping people establish their identities. (12-13)

9. Discuss how our conceptions of the various *age strata* have been shaped by cultural and historical factors. (13-15)

10. Describe the five steps employed in the scientific method. (15)
    a.

    b.

    c.

    d.

    e.

11. Compare and contrast *longitudinal* and *cross-sectional* research methods, identifying the advantages and disadvantages of each. (16-21)

12. Briefly discuss the *case-study method*. (21)

13. Summarize the following terms as they relate to the *experimental method*, and provide an example of each: (21-23)
    a. *independent variable*

    b. *dependent variable*

    c. *control group*

    d. *experimental group*

14. Define the following research designs, and critically examine the advantages and disadvantages of each: (24-27)
    a. *social survey method*

    b. *naturalistic observation method*

    c. *cross-cultural method*

15. As a researcher, list five tips from the fourteen tips for observing children in the lab or in the field. (Box, 25)

a.

b.

c.

d.

e.

16. Identify the major ethical concerns that are incorporated into guidelines for doing research with human subjects. (27-28)

KEY TERMS

---

age cohort (page 10)

age strata (13)

ascription (12)

case-study method (21)

cognitive development (6)

control group (23)

cross-cultural method (26)

cross-sectional method (19)

culture (13)

dependent variable (22)

5

development (4)

ecological approach (8)

event sampling (26)

experimental group (23)

experimental method (21)

growth (6)

independent variable (22)

learning (7)

longitudinal method (16)

maturation (6)

naturalistic observation (24)

physical development (6)

psychosocial development (6)

survey (24)

time sampling (26)

Multiple-Choice

Circle the letter of the response which *best* completes or answers each of the following statements and questions.

Factual Questions:

1. Psychologists who study orderly and sequential changes that occur in behavior with the passage of time are studying

   a. growth
   b. maturation
   c. development
   d. learning

2. The four major objectives of developmental psychology are describing, explaining, predicting, and _____ developmental changes.

   a. modifying
   b. redirecting
   c. changing
   d. controlling

3. Developmental psychologists who study psychosocial development are interested in

   a. perception
   b. mental activity
   c. personality formation
   d. motor skills

4. When a particular biological potential, such as the ability to walk, automatically unfolds in a set, irreversible sequence, we refer to this process as

   a. growth
   b. maturation
   c. learning
   d. development

5. An advocate of the ecological approach believes that the study of developmental influences must include

   a. a person's interaction with the environment
   b. the individual's changing physical and social settings
   c. how the process is affected by society
   d. all of the above

6. An example of a normative age-graded influence on development would be an adolescent

   a. experiencing a sudden religious conversion
   b. in the 1930s dropping out of school
   c. developing a severe case of acne
   d. finishing junior high school

7

7. When a developmental psychologist says that *all societies* use age for ascription, she means that

   a. roles are assigned independent of the person's unique abilities
   b. ageism is a form of prejudice in some societies
   c. as a person ages, he or she is afforded more societal status
   d. young men are assigned to the military in some cultures

8. The historian deMause has asserted that

   a. throughout history, childhood has been considered a distinct age stage
   b. the idea of children needing continuous guidance and support developed in the nineteenth century
   c. the media's depiction of such issues as violence, sex, and illness deprives children of childhood
   d. the distinction between the young-old and the old-old is not conceptually useful

9. A hypothesis refers to a(n)

   a. prediction that can be tested by gathering appropriate information
   b. controlled experimentation conducted in a lab setting
   c. explanation of experimental data
   d. test to determine whether a prediction is correct

10. In the cross-sectional method

   a. the same group of subjects is repeatedly given the same test over a twenty-year period
   b. surveys are administered to samples of people from around the country
   c. different groups of children are observed at the same time
   d. the behavior of subjects in a laboratory environment is compared with their behavior in their natural setting

11. One of the major criticisms of cross-sectional studies is that

   a. they are costly and time-consuming to conduct
   b. it is difficult to keep in contact with all the subjects
   c. it is difficult to control the environment of the subjects between testing periods
   d. the comparability of the groups being tested is difficult to assess

12. The case-study approach

   a. relies mainly on the use of surveys
   b. is exemplified by the baby biography
   c. focuses on getting a representative sample of children
   d. is used to compare different individuals systematically

13. In an experiment, the dependent variable is

   a. a measure of behavior
   b. the variable being manipulated
   c. administered only to the experimental group
   d. administered only to the control group

14. A developmental psychologist tries to determine whether a variable (e.g., instructional method) that she systematically manipulates affects another variable (e.g., IQ scores). This psychologist is using the _____ method.

   a. correlational
   b. experimental
   c. social survey
   d. case study

15. An important ethical guideline for human research that was emphasized by the author of your text was that

   a. subjects should be required to continue in a study once they commit themselves
   b. subjects must sometimes be coerced to participate in an experiment
   c. the experimenter is responsible for correcting any undesirable results stemming from subjects' participation
   d. deception should be used when full disclosure to subjects would be harmful to their egos

Conceptual Questions:

1. On the basis of previous research, a pediatrician decides to place children diagnosed with attention deficit disorder on a medication regimen including the drug Ritalin. This action is an example of

   a. control over behavioral development
   b. prediction of developmental change
   c. explanation of developmental change
   d. an unethical practice

2. A scientist wishes to *explain* the specific determinants of a child's change in language. Which statement best represents the type of question she might ask when trying to *explain* this change?

   a. What is the role of learning in language acquisition?
   b. What is the nature of the speech?
   c. What are the language capabilities of a 6-month-old infant likely to be at 14 months of age?
   d. Can we control this change in language?

3. A psychologist, conducting research on the memory strategies used by 7-year-old children, is focusing on

   a. maturational development
   b. affective development
   c. psychosocial development
   d. cognitive development

4. A developmental psychologist discovers that young women begin menstruating only after they have reached a certain weight, body fat proportion, and height and after other sexual changes have occurred. This psychologist might reasonably conclude that menstruation is controlled by

   a. learned readiness
   b. growth gradients
   c. maturational factors
   d. ecological factors

5.   Which of the following scenarios illustrates an individual who is experiencing a developmental change in *growth*?

a.   A 7-month-old infant sits upright without support.
b.   An adolescent grows 6 inches taller over the course of a year.
c.   A young child decides not to touch the hot stove because he was burned when he did this before.
d.   A 12-year-old girl begins to ovulate.

6.   Unlike American families 100 years ago, in many families today the mother works outside the home, and close family relatives may be separated by substantial geographical distances. Developmental psychologists who view development from an ecological approach would explain these changes by referring to the _____ within which today's families exist.

a.   mesosystem
b.   exosystem
c.   macrosystem
d.   microsystem

7.   Which of the following best exemplifies age stratification?

a.   A lawyer is granted a personal parking space after being promoted to district attorney.
b.   Adolescents in all cultures reach puberty at relatively the same age.
c.   A 70-year-old man is criticized by his friends for marrying a much younger woman.
d.   At 16, a young woman passes her driving test.

8.   Louis Terman did a study in which the intelligence test performance of the same group of children was assessed at different points in their lifetimes. This study was a

a.   sequential study
b.   longitudinal study
c.   cross-sectional study
d.   time-sampling study

9.   A psychologist does a study in which he compares the IQ test performance of groups of 30-, 40-, and 50-year-old people. Each group's average score is lower than the preceding group's scores. The psychologist concludes that as people age, their test scores decline. This conclusion may be in error because

a.   this is a longitudinal study
b.   IQ tests are not valid measures of intelligence
c.   you cannot give the same IQ tests to people who vary in age
d.   the different age groups differed not only in age but also in cohort

10. A psychologist studies the emotional development of her son by recording the different ages at which different emotions appear. This research is an example of a(n)

a.   case study
b.   experiment
c.   naturalistic observation
d.   cross-sectional method

11. A group of students in a psychology class conduct an experiment in which the influence of an observer's presence on the subject's willingness to help is measured. The observer's presence is the

    a. dependent variable
    b. independent variable
    c. controlled variable
    d. varied variable

12. An experimenter is testing the effects of observed violence on children's behavior. One group of children views a violent cartoon. A second group views a humorous nonviolent cartoon. The third group is not exposed to any cartoon. The first group is a(n)

    a. experimental group
    b. control group
    c. comparison group
    d. observational group

13. A social psychologist is interested in measuring the attitudes of a group of high school teachers toward including intellectually and emotionally challenged learners in their classrooms. Which type of study would she be most likely to conduct?

    a. an experiment
    b. a naturalistic observation
    c. a survey
    d. a time-series study

14. A researcher did a study of children's prosocial behavior. He recorded the incidents in which children responded positively to another child in a preschool setting. The researcher was performing a(n)

    a. experiment
    b. longitudinal study
    c. case study
    d. naturalistic observation

15. A student in an elementary classroom realizes that he is being observed by the school psychologist. The psychologist's observations of the child's behavior may be biased by the

    a. lack of a control group
    b. Hawthorne effect
    c. lack of informed consent of the child
    d. lack of a dependent variable manipulation

16. A developmental psychologist is interested in comparing the attachment level to the parents of children raised on a collective Israel kibbutz with that of children reared in American farm families. This psychologist is using the

    a. cross-sectional method
    b. cross-cultural method
    c. naturalistic observation method
    d. cultural correlation method

Matching: Definitions

Match the letter of the term in the box with its definition.

a.	age cohort	i.	experimental
b.	ascription	j.	maturation
c.	culture	k.	growth
d.	independent	l.	dependent variable
e.	control	m.	cognitive
f.	development	n.	ecological
g.	time sampling	o.	psychosocial
h.	learning	p.	event sampling

1. happens when people modify their behavior on the basis of experience _____

2. a group whose members are of the same generation _____

3. a factor affected by experimental manipulation _____

4. orderly, sequential changes occurring in an organism _____

5. involves using age for assigning people roles without regard for unique abilities _____

6. records time taken up by an episode of behavior _____

7. automatic unfolding of biological potential _____

8. learned patterns of thinking, feeling, and acting _____

9. variable manipulated in an experiment _____

10. behavior frequency count in specified time intervals _____

11. aspect of development which focuses on personality _____

12. type of group which does not have change introduced into them _____

13. approach emphasizes influences of environmental contexts on development _____

14. group into which independent variable is introduced _____

15. aspect of development focusing on mental ability _____

16. associated with progressive increments in size with advancing age _____

Matching: Research Methods

Match one of the boldly lettered research methods in the box with each of the appropriate descriptions in the list which follows.

LM Longitudinal Method   **SS Social Survey Method**

CS Cross-Sectional Method   **NO Naturalistic Observation Method**

CSM Case-Study Method   **CC Cross-Cultural Method**

EM Experimental Method

---

1.   a procedure using interviews and questionnaires ____

2.   involves the comparison of different groups of people ranging in age ____

3.   type of study which involves data comparison from two or more societies ____

4.   a particular type of longitudinal study which focuses on one individual ____

5.   a type of study in which the same individuals are observed at different points in their lives ____

6.   involves watching and recording behavior as it occurs ____

7.   involves manipulating variables and measuring the resulting behavior change ____

8.   does not allow for systematically comparing people ____

9.   an advantage of this type of developmental research is that it is independent of the subject's ability or willingness to report on given matters ____

10.   one of the most rigorously objective research techniques available to science ____

13

Completion

Supply the term or terms needed to complete each of the following statements.

1. _____ over the life span is a process of becoming something different while remaining in some respects the same.

2. _____ could not take place without growth and maturation.

3. According to Bronfenbrenner, the _____ is the environmental context which consists of the interrelationships among the various settings in which the developing person is immersed.

4. The transition from worker to retiree would be considered a _____ influence.

5. _____ are social layers that are based on time periods in life.

6. One difficulty of cross-sectional research on intelligence is that it does not make allowance for _____ differences in performance.

7. Jean Piaget's intensive studies of his own children's cognitive development are examples of the _____ method.

8. In an experiment designed to determine whether the whole-word method or the phonics method improves children's reading scores, the method of instruction would be the _____ variable.

9. A psychologist uses questionnaires to assess the attitudes of college students toward the use of contraceptives. She is using the _____ research method.

10. Research suggests that in many cases children over _____ years of age are able to make sensible decisions about whether or not to take part in research.

Multiple-Choice

Factual

1.	c	(4)
2.	d	(5)
3.	c	(6)
4.	b	(6)
5.	d	(8)
6.	d	(10)
7.	a	(12)
8.	b	(14)
9.	a	(15)
10.	c	(19)
11.	d	(20-21)
12.	b	(21)
13.	a	(22)
14.	b	(21-22)
15.	c	(28)

Conceptual

1.	a	(5)
2.	a	(5)
3.	d	(6)
4.	c	(6-7)
5.	b	(6)
6.	b	(9)
7.	c	(13-14)
8.	b	(16-17)
9.	d	(20-21)
10.	a	(21)
11.	b	(22)
12.	a	(23)
13.	c	(24)
14.	d	(24)
15.	b	(26)
16.	b	(26)

Matching: Definitions

1.	h	(7)		9.	d	(22)
2.	a	(10-11)		10.	g	(26)
3.	l	(22)		11.	o	(6)
4.	f	(4)		12.	e	(23)
5.	b	(12)		13.	n	(8)
6.	p	(26)		14.	i	(23)
7.	j	(6)		15.	m	(6)
8.	c	(13)		16.	k	(6)

Matching: Research Methods

1.	SS	(24)		6.	NO	(24)
2.	CS	(19)		7.	EM	(21)
3.	CC	(26)		8.	CSM	(21)
4.	CSM	(21)		9.	NO	(24)
5.	LM	(16)		10.	EM	(21)

Completion

1.	human development (4)		6.	generational (20-21)
2.	learning (7)		7.	case-study method (21)
3.	mesosystem (8)		8.	independent (22)
4.	normative age-graded (10)		9.	social survey (24)
5.	age strata (13)		10.	9 (28)

Chapter 2

Developmental Theories

INTRODUCTION

Chapter 1 explored the subject of human development and critiqued several research methods used in the study of various aspects of human development. Chapter 2 provides an in-depth analysis of the five major theoretical perspectives in regard to human development. The chapter presents a detailed discussion of the following theories:

- Psychoanalytic Theory. This illustrates the importance of early experience in personality development and the role of unconscious motivation. Freud's psychosexual stages, as well as Erikson's psychosocial stages of development, are examined within this context.

- Behavioral Theory. This stresses the part that learning plays in inducing individuals to act in the ways that they do. Skinner's basic tenets in regard to this perspective are highlighted.

- Humanistic Theory. This emphasizes that people are capable of intervening in the flow of life's experiences to influence and fashion their own beings. Discussion centers on the viewpoints of Maslow and Rogers.

- Cognitive Theory. This draws our attention to the importance of various mental capabilities and problem-solving skills that equip people with potent adaptive skills. Piaget's cognitive stages are examined within this context.

- Evolutionary Adaptation Theory. This highlights various biological patterns that ready individuals for specific kinds of behavior.

- Other Theories. Finally, a variety of models on human development are presented to increase our understanding of the major theories.

   I. Psychoanalytic Theories

       A. Sigmund Freud: Psychosexual Stages of Development

       B. Erik Erikson: Psychosocial Stages of Development

  II. Behavioral Theory

 III. Humanistic Theory

  IV. Cognitive Theory

       A. Jean Piaget: Cognitive Stages in Development

       B. Cognitive Learning and Information Processing

   V. Evolutionary Adaptation Theory

       A. Ethology

       B. Sociobiology

  VI. Theoretical Controversy

       A. Mechanistic and Organismic Models

       B. Continuity and Discontinuity in Development

LEARNING OBJECTIVES

After completing Chapter 2, you should be able to:

1. Define *theory*, and describe the functions that a theory serves. (page 34)

2. Briefly describe Freud's three key *psychosexual* stages and their association with *fixations*. (36-37)
    a.

    b.

    c.

3. List and describe Freud's defense mechanisms, and give an example of each. (38)

   a.

   b.

   c.

   d.

   e.

   f.

   g.

   h.

4. Distinguish between Erikson's *psychosocial* and Freud's *psychosexual* development theories. (36-40)

5. Describe the approximate time period during which each psychosocial *stage* occurs and the type of life crisis experienced. (40-42)

6. Evaluate the basic principles of Freud's *psychoanalytic* viewpoint and Erikson's psychosocial theory. (38-42)

7. Discuss the focus of *behavioral theory*, and explain the significance of the following terms: (43-45)

   a. *conditioning*

   b. *stimulus*

   c. *response*

   d. *reinforcement*

e. *learned helplessness*

f. *behavior modification*

8. Briefly summarize the main principles of *humanistic* theory. (45-47)

9. Characterize a *self-actualized* individual as described by Maslow. (46)

10. Briefly describe Maslow's *hierarchy of needs* that motivate human behavior. (46)

11. Explain Piaget's view of the *cognitive stages of development*. (47-48)

12. Define the roles of the following as they relate to the process of cognitive development: (48)

    a. *scheme*

    b. *assimilation*

    c. *accommodation*

    d. *equilibrium*

13. Trace the characteristics of children at each of Piaget's *stages of cognitive development*. (50)

14. Briefly explain why it has been suggested that Piaget underestimated the cognitive capabilities of infants and young children. (51)

15. Give a brief summary of the current status of Piagetian theory. (51-52)

16. Explain *cognitive learning* and why it is significant. (52-53)

17. Define the term *natural selection*. (53)

18. Describe the contribution of the *ethological* viewpoint to our understanding of cognitive development. (52-55)

19. Briefly define the following terms as they relate to *ethology*: (54)

   a. *releasing stimulus*

   b. *imprinting*

   c. *critical period*

20. Explain why some developmental psychologists prefer the term *sensitive period* instead of *critical period*. (54)

21. Briefly describe the major principles of the *sociobiological* viewpoint. (55)

22. Describe how Jerome Kagan's work has led him to conclude that children are more resilient than we think. (Box, 56-57)

23. Define the term *kin selection* as proposed by the sociobiologist Edward O. Wilson. (56)

24. Explain why critics object to the *sociobiological* viewpoint. (57-58)

25. Distinguish between the *mechanistic, organismic,* and *electic models* of human development. (58-59)

26. Explain how proponents of the continuity and discontinuity models view human development. (59-60)

KEY TERMS

accommodation (page 48)

assimilation (48)

behavior modification (44)

behavioral theory (43)

cognition (47)

cognitive learning (53)

cognitive stages in development (47)

cognitive theory (47)

conditioning (43)

critical period (54)

eclectic approach (59)

epigenetic principle (40)

equilibrium (49)

ethology (53)

fixation (36)

hierarchy of needs (46)

holistic approach (45)

humanistic psychology (45)

imprinting (54)

kin selection (56)

learned helplessness (44)

mechanistic model (58)

natural selection (53)

organismic model (59)

peak experience (46)

psychoanalytic theory (35)

psychosexual stages (36)

reinforcement (43)

releasing stimulus (54)

response (43)

scheme (48)

self-actualization (46)

sensitive period (54)

sociobiology (55)

stages (49)

stimulus (43)

theory (34)

Multiple-Choice

Circle the letter of the response which *best* completes or answers each of the following statements and questions.

Factual Questions:

1. The major function of a theory is to

    a. organize factual observations in a coherent way
    b. describe and catalog our behavior observations
    c. prove the correctness of our hypothetical formulations
    d. determine whether a behavior is innate or learned

2. Which of the following is not characteristic of a theory?

    a. It is a set of interrelated statements.
    b. It suggests that development is directed from within.
    c. It serves as a guide to action.
    d. It is an attempt to make sense of our experiences.

3. The psychologist Jerome Bruner suggests that

    a. human development can be explained by one true theory
    b. there is only one method for describing human nature
    c. a theory can be only more or less right
    d. all of the above

4. A major premise of Freudian theory is that fixation occurs when

    a. sex-role socialization takes place
    b. a person experiences a psychosocial crisis
    c. a person is frustrated or overindulged at a stage of development
    d. a biochemical imbalance exists at a stage of development

5. According to Freud, the latency period corresponds to which period of life?

    a. infancy
    b. childhood
    c. adolescence
    d. young adulthood

6. Politicians are often accused by many psychologists of using the defense mechanism of projection. This implies that the politicians

    a. behave in an infantile way when under stress
    b. attribute their own unacceptable impulses to others
    c. are hypocritical and say things inconsistent with their beliefs
    d. tend to ignore painful reminders of their misbehavior

7. In contrast to Freud, Erik Erikson emphasized

   a. psychosocial development
   b. cognitive development
   c. sexual fixations
   d. critical periods

8. Erikson indicates that individuals typically confront a crisis associated with identity vs. role confusion at which period of life?

   a. infancy
   b. childhood
   c. adolescence
   d. old age

9. In Erikson's theory, a developing person is likely to feel shame and doubt when he or she

   a. develops an ego identity
   b. feels hopeless about impending death
   c. fails to receive recognition for school achievements
   d. has parents who are overprotective

10. The primary focus of behavioral theory is

   a. an individual's unconscious conflicts
   b. the identification of critical periods of development
   c. how people learn to behave in particular situations
   d. how people learn to resolve psychosexual crises

11. Reinforcement occurs when

   a. two stimuli are paired with each other
   b. a stimulus is followed by a response
   c. one event strengthens the probability that another event will occur
   d. a response is initiated by a releasing stimulus

12. A therapist diagnoses a client as suffering from learned helplessness. Her diagnosis implies that the individual

   a. has been punished for being assertive
   b. feels that his or her responses are doomed to failure
   c. has developed a fatalistic attitude associated with terminal illness
   d. inherited a disposition toward depressed behavior

13. Which of the following behaviors is not necessarily representative of the self-actualized personality?

   a. high intellectual aptitude
   b. problem-centered
   c. autonomy and independence
   d. spontaneity in thought and behavior

14. According to Maslow, self-actualized people tend to be

   a. older
   b. motivated by their own personal needs
   c. subject to "peak experiences"
   d. in little need of privacy

15. According to Piagetian theory, when a child engages in the process of assimilation, she

    a. is rewarded for responding accurately in a learning task
    b. conceptualizes new situations as if they were similar to previously experienced ones
    c. changes her old behavior when it no longer allows her to solve problems
    d. performs a new behavior without having had a previous opportunity to make the response

16. The behavioral development that is most characteristic of the sensorimotor stage is

    a. fixation
    b. symbolic ability
    c. object permanence
    d. egocentrism

17. A mounting body of evidence suggests that Piaget underestimated the _____ development of infants and young children.

    a. psychomotor
    b. cognitive
    c. emotional
    d. psychological

18. Cognitive learning refers to

    a. insightful learning
    b. one-trial learning
    c. intentional learning
    d. observational learning

19. According to the principle of natural selection

    a. the process of imprinting occurs only during a relatively short period of time
    b. human babies are biologically preadapted with behaviors that are prompted by releasing stimuli
    c. people who are best adapted to their environment stand a better chance of surviving and reproducing
    d. the social behavior of human beings can be explained by referring to our common genetic heritage

20. In contrast to mechanistic models of human development, organismic models portray human beings as

    a. intrinsically active
    b. intrinsically passive
    c. developing in a continuous process
    d. mostly influenced by environmental factors

Conceptual Questions:

1. The adult is compulsively neat, orderly, and stingy. Freudian psychologists would say that this individual may have

    a. been weaned too early
    b. experienced a conflict related to toilet training
    c. never resolved his Oedipus conflict
    d. been fixated at the guilt stage

2. Maria acts very hostile toward others when actually she feels tender, weak, and vulnerable. According to Freud, this is an example of what?

   a. rationalization
   b. isolation
   c. reaction formation
   d. sublimation

3. John feels that going to college will be a waste of time because he was severely criticized by his parents and teachers for poor academic performance in third grade. According to Erikson's psychosocial theory, John's problems may be related to a crisis experienced at what stage?

   a. identity vs. role confusion
   b. autonomy vs. shame and doubt
   c. industry vs. inferiority
   d. initiative vs. guilt

4. A woman becomes concerned with future generations and with others beyond the immediate family. According to Erikson, she is most likely in which stage of development?

   a. puberty
   b. young adulthood
   c. middle adulthood
   d. old age

5. A young boy has learned to associate a grimace on his mother's face with an impending crack on the behind. If he gets anxious when his mother grimaces, his anxiety is due to

   a. the conditioning of that response
   b. learned shame and doubt
   c. his mother's face being a releasing stimulus
   d. negative reinforcement

6. A fifth-grade child believes that no matter how hard she studies for a math test, she will probably fail. A learning theorist would say that the child is experiencing

   a. frustration at a psychosocial stage
   b. reaction formation
   c. learned helplessness
   d. learned behavior modification

7. The title of a song is "I Gotta Be Me." This most clearly is a direction taken by which theory?

   a. humanistic
   b. psychoanalytic
   c. ethological
   d. sociobiological

8. A person grows up in an inner-city ghetto and fails to receive proper medical care and nutrition. Humanistic psychologists would say that such an individual may

   a. develop learned helplessness
   b. be less likely to achieve self-actualization
   c. be more likely to have an air of detachment and a need for privacy
   d. have a tendency to repress memories of this painful experience

9. A child drinks a shot glass full of whiskey as if he were drinking a glass of water. The child is demonstrating the Piagetian process of

   a. equilibrium
   b. accommodation
   c. assimilation
   d. regression

10. An infant has previously been successful in getting a toy by swiping at it with her hand, but now the toy is out of reach and cannot be retrieved. Instead, the child _____ her actions and acquires the toy by tugging on the edge of a tablecloth that the toy rests on.

    a. assimilates
    b. accommodates
    c. behaviorally modifies
    d. equilibrates

11. Michael realizes that when water is poured from a tall thin glass into a short wide glass, the same amount of water exists. Piaget would argue that the child

    a. achieved the concept of conservation through observational learning
    b. is demonstrating abstract thought
    c. is demonstrating object permanence
    d. is exhibiting egocentrism

12. A child can respond correctly to the following problem: If ice heated things, we could boil water by putting it in the refrigerator. This child is in the

    a. anal stage
    b. stage of initiative
    c. stage of concrete operations
    d. formal operations stage

13. Dr. Perez argues that intelligent people have a better chance of surviving and that therefore the most prosperous countries are populated by a greater percentage of bright people. She is focusing on the effects of

    a. evolutionary progression
    b. native endowment
    c. natural selection
    d. imprinting

14. A newborn calf is separated from its mother for three hours following birth. Subsequently, the cow rejects its newborn and fails to feed it. This demonstrates the importance of

    a. observational learning
    b. critical periods
    c. natural selection
    d. conditioning

15. A developmental psychologist asserts that males are not naturally monogamous because if they were, they would decrease the chances of perpetuating their gene pool and passing on their social characteristics to subsequent generations. This psychologist would probably be characterized as a

    a. cognitivist
    b. psychoanalyst
    c. humanist
    d. sociobiologist

Matching: Definitions

Match the letter(s) of the term in the box with its definition.

```
a. accommodation              q. kin selection
b. cognition                  r. cognitive learning
c. equilibrium                s. epigenetic principle
d. behavior modification      t. learned helplessness
e. peak experiences           u. natural selection
f. theory                     v. stages
g. reinforcement              w. responses
h. eclectic approach          x. cognitive stages in
i. imprinting                    development
j. organismic model           y. critical period
k. assimilation               z. mechanistic model
l. hierarchy of needs         AA fixation
m. self-actualization         BB conditioning
n. stimuli                    CC releasing stimulus
o. psychosexual stages        DD schemes
p. holistic approach          EE sensitive period
```

1. expectancy that events are independent of one's responses _____

2. suggests that all lower level needs must be satisfied before self-actualization can occur _____

3. function as especially potent activators of parenting _____

4. using learning theory to change maladaptive behavior _____

5. reaction to a stimulus _____

6. sequential periods associated with the evolution of thinking _____

7. survival of individuals better adapted to their environment _____

8. view of the organism as an active being and as an organized configuration different from its parts _____

9. balance between assimilation and accommodation _____

10. each facet of development must occur at a particular time; otherwise, subsequent development is altered _____

11. organizational framework for summarizing and explaining factual observations _____

12. cognitive structure used for dealing with specific environmental situations _____

13. according to Piaget, these are age-related events experienced by children _____

14. processes such as perception and retention _____

15. portrayal of humans as reactive and of development as gradual, uninterrupted, and chainlike

_____

16. environmental units which provoke responses _____

17. a type of associative learning _____

18. stages of Freudian theory _____

19. one event's strengthening the possibility of another event's occurrence _____

20. time during development when specific experiences have their greatest impact _____

21. one who has developed his or her potential to the fullest _____

22. developmental arrest due to overindulgence or frustration, creating characteristic adult behaviors _____

23. interpreting new information so that it conforms with a currently held schema _____

24. excitement, insight, and happiness experienced by self-actualized people _____

25. preferred term for critical periods in human development _____

26. observational or social learning _____

27. choosing the best aspects of each theory for description and analysis _____

28. view of human beings as more than a collection of physical and psychological components _____

29. changing a schema to achieve a better match with reality _____

30. innate attachment process that occurs at a particular time _____

31. evolution favors genes that improve the chances of a group's survival _____

Matching: Theoretical Perspectives on Development

Match each of the lettered items in the box with the appropriate item or items in the list that follows.

PA Psychoanalytic Theory   C   Cognitive Theory

S   Sociobiology   E   Ethology

H   Humanistic Theory   PS Psychosocial Theory

B   Behavioral Theory

---

1. focuses on personality development through stages and unconscious motivation _____

2. an approach focusing on how one learns to behave _____

3. stresses human uniqueness and active control of one's destiny _____

4. focuses on the biological foundations of social behavior _____

5. suggests that organisms are genetically prepared for some responses _____

6. a theorist from this viewpoint would be interested only in collecting observable and measurable data _____

7. focuses on how organisms receive, store, transform, and use information _____

8. views development as a series of stages which confront individuals with a *crisis* that they must struggle through _____

9. according to this perspective, the key to life is survival _____

10. suggests that people must resolve *conflicts* at various stages of development _____

11. views behavior as part of the adaptational package of an organism and as necessary for its survival _____

12. this perspective holds that people are self-determining _____

13. views people as information processors _____

14. according to this theory, development follows the epigenetic principle _____

Completion

Supply the term or terms needed to complete each of the following statements.

1. A _____ helps us to summarize, organize, and explain our observations concerning human development.

2. A person who becomes _____ at the oral psychosexual stage can be described as immature and dependent.

3. Someone who represses unacceptable behavior and replaces it with socially acceptable behavior is using the defense mechanism of _____

4. A child who experiences the life crisis of _____ vs. _____ is affected by the amount of freedom his parents provide him during the preschool years.

5. A teacher decides to reduce the frequency of a student's out-of-seat behavior by ignoring (not reinforcing with attention) the child when he is out of his seat. This teacher is using _____, or applied learning theory.

6. Humanistic psychologists might describe a _____ _____ person as self-accepting, spontaneous, problem-centered, and autonomous.

7. A child calls a tiger at the zoo a "kitty." After being corrected, the child uses the term tiger. Piaget would assert that the child had changed her _____ through accommodation, so that it might better match the world of reality.

8. When infants do not realize that objects have an independent existence, Piaget asserts that these infants lack object _____.

9. _____ learning theorists emphasize the value of our symbol-using capacity in allowing us to comprehend and deal with the environment.

10. Ethologists assert that the large head, small body, and distinctive features of small infants serve as _____ _____ for eliciting nurturing responses from their adult caretakers.

11. Developmental psychologists who accept the _____ model of human development view humans as reactive organisms.

ANSWERS FOR SELF-TESTS - CHAPTER 2

Multiple-Choice

Factual				Conceptual		
1.	a	(34)		1.	b	(37)
2.	b	(34)		2.	c	(38)
3.	c	(34)		3.	c	(41)
4.	c	(36)		4.	c	(42)
5.	b	(36)		5.	a	(43)
6.	b	(38)		6.	c	(44)
7.	a	(39)		7.	a	(45)
8.	c	(41)		8.	b	(46)
9.	d	(41)		9.	c	(48)
10.	c	(43)		10.	b	(48-49)
11.	c	(43)		11.	a	(50)
12.	b	(44)		12.	d	(50-51)
13.	a	(46)		13.	c	(53)
14.	c	(46)		14.	b	(54)
15.	b	(48)		15.	d	(55)
16.	c	(50)				
17.	b	(51)				
18.	d	(53)				
19.	c	(53)				
20.	a	(59)				

Matching: Definitions

1. t   (44)
2. l   (46)
3. CC   (54)
4. d   (44)
5. w   (43)
6. x   (47)
7. u   (53)
8. j   (59)
9. c   (49)
10. s   (40)
11. f   (34)
12. DD   (48)
13. v   (49)
14. b   (47)
15. z   (58)

16. n   (43)
17. BB   (43)
18. o   (36)
19. g   (43)
20. y   (54)
21. m   (46)
22. AA   (36)
23. k   (48)
24. e   (46)
25. EE   (54)
26. r   (53)
27. h   (59)
28. p   (45)
29. a   (48-49)
30. i   (54)
31. q   (56)

Matching: Theoretical Perspectives

1. PA   (35-36)
2. B   (43)
3. H   (45)
4. S   (55)
5. E   (53)
6. B   (43)
7. C   (47)

8. PS   (39)
9. S   (55)
10. PA   (35-36)
11. E   (53)
12. H   (45)
13. C   (47)
14. PS   (39-40)

Completion

1. theory (34)
2. fixated (37)
3. sublimation (38)
4. initiative; guilt (41)
5. behavior modification (44)
6. self-actualized (46)

7. scheme (48)
8. permanence (50)
9. cognitive (53)
10. releasing stimuli (54)
11. mechanistic (58-59)

Chapter 3

Biological Foundations

INTRODUCTION

Whereas Chapters 1 and 2 examined the major developmental perspectives and theories, Chapter 3 explores the biological foundations of development. The chapter presents what the most current research has found about the beginning of human development. Chapter 3 examines such important topics as reproduction, genetics, and the dynamic interaction between heredity and environment. Within this context, the following topics are covered:

- Infertility. How this problem is being treated is examined, as well as a discussion on the issue of surrogate parenthood.

- Genetics. An overview is presented of the basic genetic principles that govern the mechanisms of heredity.⊠

- Genetic Assessment and Genetic Counseling. The methods available to assess development before birth are discussed, including amniocentesis, ultrasonography, and chorionic villus biopsy.

- Hereditary and Environmental Factors. The chapter examines how specific factors combine to influence the developing organism.

⊠ This symbol indicates that relevant material has been placed in Appendix A: Genetics and Heredity, beginning on p. 611.

33

I.   Reproduction

    A.   Reproductive Systems

    B.   Fertilization

    C.   Infertility and Its Treatment

    D.   Genetic Counseling and Testing

II.   Genetics

    A.   Chromosomes ⊠

    B.   Genes ⊠

    C.   Dominance ⊠

    D.   Mendel's Principles ⊠

    E.   Determination of Sex ⊠

    F.   Linked Characteristics ⊠

    G.   Chromosomal Abnormalities ⊠

    H.   Incomplete Dominance ⊠

    I.   Penetrance and the Norm of Reaction ⊠

III.   Heredity-Environment Interactions

    A.   The "Which" Question

    B.   The "How Much" Question

    C.   The "How" Question

LEARNING OBJECTIVES

After completing Chapter 3, you should be able to:

1. Distinguish between male and female sex cells and between male and female reproductive organs. (page 66)

2. Describe the *fertilization* process that may occur during a normal menstrual cycle. (68-69)

3. Briefly discuss fertility peaks as they relate to human beings. (69-70)

4. Define *infertility*, discuss its incidence rate among U.S. couples, identify contributing factors, and cite the various techniques available to treat this problem. (70-75)

5. Discuss the ethical issues raised by surrogate motherhood. (74-75)

6. Explain the following terms: (⊠ 612-615)

   a. genetics

   b. chromosomes

   c. mitosis

   d. genes

   e. DNA

   f. allele

7. Differentiate between the following: (⊠ 614-615)

   a. dominant and recessive

   b. genotype and phenotype

   c. homozygous and heterozygous

8. Briefly explain Mendel's principles of segregation and independent assortment, and identify one dominant gene and one recessive-gene disorder. (⊠ 615)

9. Distinguish between autosomes and sex chromosomes. (⊠ 616-618)

10. Identify three sex-linked disorders. (⊠ 617)

11. Discuss four different chromosomal abnormalities. (⊠ 618-619)

12. Provide examples of incomplete dominance, penetrance, and the norm of reaction. (⊠ 619-620)

13. Describe how *amniocentesis, ultrasonography,* and *chorionic villus biopsy* assist in the process of genetic counseling and identify the most recent research relevant to each technique. (73-76)

14. Briefly explain how genetic screening could lead to new ways for people to discriminate. (77)

15. Discuss and critically evaluate the "which," "how much," and "how" models in the nature-nurture controversy.   (77-85)

16. Develop an opinion regarding the validity of family resemblance studies and the usefulness of behavioral research. (78-79)

17. Cite research to support the notion that children should be treated individually. (84-85)

18. Briefly discuss how research with animals has provided insight into how *heredity* and *environmental* factors interact to shape development. (84)

19. List several characteristics of those "superchildren" who thrive despite enormous difficulties. (85)

KEY TERMS

---

amniocentesis (page 73)

chorionic villus biopsy (76)

continuum of indirectness (83)

37

environment (78)

fertilization (69)

fraternal (dizygotic) twins (79)

heredity (78)

identical (monozygotic) twins (79)

infertility (70)

in vitro fertilization (71-72)

polygenic inheritance (81)

reproduction (66)

ultrasonography (76)

SELF-TESTS   CHAPTER 3 & APPENDIX A

Multiple-Choice

Circle the letter of the response which *best* completes or answers each of the following statements and questions.

Factual Questions:

1. Sex cells (sperm and ova) are called

    a. alleles
    b. genes
    c. gametes
    d. zygotes

2. When the Graafian follicle ruptures and releases an ovum, it transforms itself into the

    a. corpus luteum
    b. epididymis
    c. gamete
    d. oviduct

3. Research has shown that human reproduction peaks during which month?

    a. April
    b. March
    c. July
    d. August

4. The most common cause of female infertility is

    a. in vitro fertilization
    b. blocked oviducts
    c. congenital or infectious problems
    d. exposure to environmental toxins

5. The biggest barrier to successful in vitro fertilization is

    a. the sperm and ovum fail to join together
    b. the drug used to stimulate production of several ova often fails to work
    c. the embryos fail to implant
    d. low success rate of the hormone treatment

6. In the absence of clear guidelines provided by legislation in regard to surrogate parenthood, judges have attempted to base their decisions on

    a. behalf of the genetic parents
    b. the best interests of the surrogate mother
    c. their own interpretation of the law
    d. the best interests of the child

7. Which statement about chromosomes is *true*? ⊠

    a. The exact number of chromosomes is the same for all species of animals.
    b. The exact nature of chromosomes is the same for all species of animals.
    c. Every cell in the human body contains the same number of chromosomes.
    d. They are located in the cell nucleus.

8. Each sperm or ovum receives only one of each kind of chromosome from every parental cell. This is half the usual number of chromosomes found in body cells, allowing each parent to contribute half of the total number of chromosomes. The process by which this reduction is achieved is called ⊠

    a. mitosis
    b. ovulation
    c. meiosis
    d. fertilization

9. Each member of a pair of genes is called a(n) ⊠
    a. autosome
    b. chromosome
    c. allele
    d. gamete

10. Gregor Mendel pointed out that there is a distinction between the genetic makeup of an organism and its observable characteristics. The observable characteristics of an organism are known as its ⊠

    a.   hybrid
    b.   genotype
    c.   phenotype
    d.   homozygous

11. According to the genetic principle of segregation, ⊠

    a.   genetic characteristics do not blend with or contaminate one another
    b.   all characteristics are inherited independently of all other characteristics
    c.   two genetic traits will blend with one another to form a hybrid
    d.   two genetic traits on the same chromosome will not be passed along together

12. If you carry a recessive gene for albinism, then your children ⊠

    a.   will be albinos if you are homozygous
    b.   will be albinos if your mate is homozygous
    c.   may be albinos if your mate has the gene too
    d.   Two genetic traits on the same chromosome will not be passed along together.

13. Which of the following is an example of a sex-linked trait? ⊠

    a.   the albino condition
    b.   hemophilia
    c.   Huntington's chorea
    d.   Joseph's disease

14. Klinefelter's syndrome, a genetic disorder, is associated with the following chromosomal condition: ⊠

    a.   XY
    b.   XXY
    c.   X
    d.   XYY

15. According to your text, an example of incomplete dominance is ⊠

    a.   Turner's syndrome
    b.   trisomy 21
    c.   Hunter's disease
    d.   sickle-cell anemia

16. According to the norm of _____, a person with a given genotype may develop different phenotypes depending on environmental conditions. ⊠
    a.   reaction
    b.   penetrance
    c.   dominance
    d.   inheritance

17. A medical procedure that makes it possible to detect whether an unborn baby will have genetic defects is called
    a. progesterone
    b. geneotosis
    c. chorionic heterozygosis
    d. amniocentesis

18. Parents who have a strong history of genetic disease in their families may consider using _____ to determine if their fetus is normal.
    a. chorionic villus biopsy
    b. behavior genetics
    c. polygenic inheritance studies
    d. qualitative indirectness studies

19. Which technique would scientists use if they wish to determine how much of a given trait is due to heredity and how much is due to environment?
    a. qualitative analysis studies
    b. family resemblance studies
    c. polygenic inheritance studies
    d. qualitative indirectness studies

20. According to the concept of a "continuum of indirectness" formulated by Anastasi, the contributions of
    a. heredity are direct
    b. environment are indirect
    c. heredity and environment are both indirect
    d. heredity to social stereotypes are indirect

21. As formulated by Anne Anastasi and others, the "how" question in the nature-nurture controversy is concerned with
    a. the respective contributions that heredity and environment make to a given characteristic
    b. the interaction that takes place between heredity and environment
    c. determining whether heredity or environment is more important
    d. how much hereditary and environmental influence is required to produce a given trait

Conceptual Questions:

1. Ovum is to _____ as sperm is to _____.
    a. gamete; zygote
    b. estrogen; androgen
    c. prostate gland; uterus
    d. fertilization; gamete

2. A couple is having trouble conceiving. The male's sperm count is normal. Of the following factors, which would most likely account for the problem?

    a. The sperm all contain twenty-three chromosomes.
    b. The ovaries are releasing too many ova.
    c. There is uterine tissue in the woman's abdominal cavity.
    d. The male has endometriosis.

3. In Aldous Huxley's novel, *Brave New World*, alpha, beta, and gamma babies are created through a government eugenics program. The practice of _____ raises similar ethical questions today.

   a. in vitro fertilization
   b. amniocentesis
   c. independent assortment
   d. Mendelian segregation

4. Meiosis is to _____ as mitosis is to body cells. ⊠
   a. zygote
   b. gametes
   c. genes
   d. autosomes

5. Both your parents have blue eyes, but your eyes are brown. The most logical explanation of this is that ⊠
   a. your mother had a heterozygous eye-color genotype
   b. blue is a dominant character
   c. you were probably adopted
   d. you have a heterozygous eye-color genotype

6. A curly-haired woman marries a man with straight hair. If all their children have curly hair, which of the following conclusions is correct? ⊠
   a. The gene for curly hair is recessive.
   b. The gene for straight hair is dominant.
   c. The grandchildren will have straight hair.
   d. The gene for curly hair is dominant.

7. A man whose father had diabetes does not develop diabetes himself. It is likely, however, that the man has ⊠
   a. a diabetic phenotype
   b. a diabetic genotype
   c. homozygous alleles for diabetes
   d. an identical genotype and phenotype

8. If all your children have dark hair but all your grandchildren have light or red hair, which of the following principles is demonstrated? ⊠
   a. multifactorial inheritance
   b. independent assortment
   c. mixed dominance
   d. segregation

9. An adult male is suffering from an autosomal dominant disease in which he experiences mental deterioration and uncontrolled twitching. It is also likely that ⊠

   a. three-quarters of his children will get the disease
   b. he has an enzyme defect
   c. he has trisomy 21
   d. his children will have a 50-50 chance of getting the disease

10. The last czarevitch of Russia suffered from hemophilia. Which of the following conclusions, therefore, would be most likely? ⊠
    a. The czarevitch's mother had hemophilia.
    b. The czarevitch's father had hemophilia and his mother was normal.
    c. The czarevitch's mother was an unaffected carrier of hemophilia.
    d. The czarevitch's father was an unaffected carrier of hemophilia.

11. You are examining the chromosomal structure of a child who has hemophilia. Which of the following structures would you most expect to find? ⊠
    a. XX
    b. YY
    c. Yy
    d. XY

12. It is determined that a young male is color blind. He probably inherited this trait from ⊠
    a. his mother
    b. his father
    c. both parents
    d. either parent; it is impossible to tell

13. A man has underdeveloped testicles, is tall and thin, and has enlarged breasts. Which of the following is probably true? ⊠
    a. He has one extra X chromosome.
    b. He has one missing Y chromosome.
    c. He has one extra autosome.
    d. He has one missing X chromosome.

14. Two identical twins are separated at birth. The one who is raised in an educationally enriched environment develops normal intelligence, whereas the one who was raised by illiterate, abusive parents develops borderline mental abilities. This result can best be explained by referring to ⊠
    a. incomplete dominance
    b. independent segregation
    c. penetrance
    d. the norm of reaction

15. A woman's obstetrician suggests to her that he needs to get a sample of amniotic fluid from her fetus. The obstetrician
    a. is performing in vitro fertilization
    b. intends to inseminate a surrogate mother
    c. may suspect that the fetus is genetically defective
    d. is conducting a fetal villus biopsy

16. Dr. Lorenz conducted a study and hypothesized that genetic factors could account for 80 percent of the observed differences in tested intelligence scores of groups of subjects. Lorenz was responding to
    a. the "which" question
    b. the "how much" question
    c. the "how" question
    d. the concordance question

17. A behavioral genetic researcher compares the IQ's of adopted children with those of their adoptive parents and their biological parents. This researcher is probably trying to answer the _____ question.

    a. "how much"
    b. "which"
    c. "how"
    d. "interaction"

18. A third-grade boy frequently searches for playmates and sometimes creates imaginary playmates if real ones are not at hand. According to your text, what type of relationship does he have with his environment?
    a. passive
    b. active
    c. reactive
    d. evocative

Matching: Definitions

Match the letter of the term in the box with its definition.

a.	reproduction	f.	fertilization
b.	environment	g.	polygenic inheritance
c.	in vitro fertilization	h.	infertility
d.	fraternal (dizygotic) twins	i.	heredity
e.	continuum of indirectness	j.	identical (monozygotic) twins

1. the union of a sperm and an ovum _____

2. process by which organisms create more organisms of their own kind _____

3. failure to achieve pregnancy after repeated efforts without contraception _____

4. twins who come from two separate eggs _____

5. biological transmission of traits through genes _____

6. procedure combining laboratory conception with subsequent natural embryonic and fetal development _____

7. traits that derive from the interaction of several genes _____

8. the view that heredity plays a more central role in some aspects of development than in others _____

9. twins who come from the same egg _____

10. external factors affecting an organism _____

Matching: Genetics (Most of this material is in Appendix A: Genetics and Heredity)

a.	homozygous	l.	genetics
b.	penetrance	m.	sex chromosomes
c.	chromosomes	n.	meiosis
d.	dominant character	o.	genotype
e.	genes	p.	amniocentesis
f.	autosomes	q.	allele
g.	principle of independent assortment	r.	incomplete dominance
h.	phenotype	s.	recessive character
i.	chorionic villus biopsy	t.	heterozygous
j.	ultrasonography	u.	norm of reaction
k.	deoxyribonucleic acid (DNA)	v.	principle of segregation

1. two cell divisions, during which the chromosome number is halved _____

2. the 22 pairs of chromosomes that are similar in size and shape in both men and women _____

3. one of two or more alternative forms of a gene _____

4. threadlike bodies which contain hereditary materials _____

5. the genetic makeup of an organism _____

6. genotype with dissimilar alleles _____

7. units of a chromosome _____

8. fetal cell analysis used to detect birth defects _____

9. the range of possible outcomes that genetic influences have in different environmental settings _____

10. the ability of a gene to manifest itself in the person who carries it _____

11. character that is masked by another allele in the phenotype _____

12. Mendel's law stating that every characteristic is inherited independently of every other characteristic _____

13. procedure used to trace size and shape of fetus _____

14. a genotype with identical alleles _____

15. a condition in which one gene is not dominant over another _____

16. technique which analyzes a sample of the tissue from the membrane around the embryo _____

17. the scientific study of biological inheritance _____

18. chromosomes that are either XX or XY _____

19. view that characteristics preserve their integrity as they pass to succeeding generations _____

20. character that masks another allele in the phenotype expression _____

21. active substance of genes telling cells how to operate _____

22. observable characteristics of an organism _____

Completion

Supply the term or terms needed to complete each of the following statements.

1. A male sperm fuses with a female ovum to form a _____.

2. In vitro fertilization is successful only _____ percent of the time.

3. If your father gave you a gene for straight hair but you have curly hair, we could say that your genotype is _____. ⊠

4. Males with _____ syndrome have two X and one Y chromosome. ⊠

5. Techniques used to study the hereditary aspects of behavior are referred to as _____ genetics.

6. An example of a(n) _____ gene environment interaction would be when a pair of socially adept parents provide their child with an enriched social setting.

7. One feature that stands out in regard to characteristics of "superchildren" is that they need _____ to become invulnerable.

8. In _____ each single chromosome splits lengthwise to form a new pair.

Multiple-Choice

Factual

1.	c	(66)
2.	a	(69)
3.	b	(69)
4.	c	(71)
5.	c	(72)
6.	d	(74)
7.	d	(⊠ 612)
8.	c	(⊠ 612))
9.	c	(⊠ 614)
10.	c	(⊠ 614)
11.	a	(⊠ 615)
12.	c	(⊠ 614-615)
13.	b	(⊠ 617)
14.	b	(⊠ 619)
15.	d	(⊠ 619)
16.	a	(⊠ 620)
17.	d	(73)
18.	a	(76)
19.	b	(78)
20.	d	(83)
21.	b	(83)

Conceptual

1.	b	(67)
2.	c	(71)
3.	a	(72-73)
4.	b	(⊠ 612)
5.	c	(⊠ 614)
6.	d	(⊠ 614
7.	b	(⊠ 614)
8.	d	(⊠ 615)
9.	d	(⊠ 616)
10.	c	(⊠ 618)
11.	d	(⊠ 618)
12.	a	(⊠ 618)
13.	a	(⊠ 619)
14.	d	(⊠ 620)
15.	c	(73)
16.	b	(78-79)
17.	a	(79)
18.	b	(85)

Matching: Definitions

1.	f	(69)
2.	a	(66)
3.	h	(70)
4.	d	(79)
5.	i	(77-78)
6.	c	(71-72)
7.	g	(81)
8.	e	(83)
9.	j	(79)
10.	b	(78)

Matching: Genetics (Most of this material is in Appendix A: Genetics and Heredity)

1.	n	(⊠ 612)
2.	f	(⊠ 616)
3.	q	(⊠ 614)
4.	c	(⊠ 612)
5.	o	(⊠ 614)
6.	t	(⊠ 614)
7.	e	(⊠ 613)
8.	p	(73)
9.	u	(⊠ 620)
10.	b	(⊠ 620)
11.	s	(⊠ 614)
12.	g	(⊠ 616)
13.	j	(76)
14.	a	(⊠ 614)
15.	r	(⊠ 619)
16.	i	(76)
17.	l	(⊠ 612)
18.	m	(⊠ 616-618)
19.	v	(⊠ 615)
20.	d	(⊠ 614)
21.	k	(⊠ 613)
22.	h	(⊠ 614)

Completion

1.	zygote (66)	
2.	12 (72)	
3.	heterozygous (⊠ 614)	
4.	Klinefelter's (⊠ 619)	
5.	behavior (79-80)	
6.	passive (84)	
7.	challenges (85)	
8.	mitosis (⊠ 612)	
9.	Huntington's chorea (⊠ 616)	
10.	family resemblance (78-79)	

Chapter 4

Prenatal Development and Birth

INTRODUCTION

Chapter 4 illuminates the remarkable and critical period of prenatal development and the process of birth. Several important topics are discussed, including the following:

- Prenatal Development. The three stages of prenatal development--germinal period, embryonic period, and the fetal period--are examined in detail.

- Prenatal Environmental Influences. This includes a discussion of the prenatal environmental influences operating from the moment of conception.

- Birth Process. This is an in-depth look at the radical changes which occur during the birth process.

- Possible Complications. The chapter covers some of the complications which may arise during pregnancy, as well as the treatments available to help the mother and the child.

I.   Stages of Prenatal Development

    A.   The Germinal Period

    B.   The Embryonic Period

    C.   The Fetal Period

II. Prenatal Environmental Influences

    A.   Drugs and Other Chemical Agents

    B.   Maternal Infectious and Noninfectious Diseases

    C.   Maternal Sensitization: The Rh Factor

    D.   Maternal Nutrition and Prenatal Care

    E.   Maternal Stress

    F.   Maternal Age

III. Birth

    A.   The Birth Process

    B.   Natural Childbirth

    C.   Parent-Infant Bonding

    D.   Birthing Rooms and Family-Centered Hospital Care

    E.   The Baby's Birth Experience

IV. Complications of Pregnancy and Birth

    A.   Toxemia

    B.   Caesarean Sections

    C.   Prematurity

LEARNING OBJECTIVES

After completing Chapter 4, you should be able to:

1.   Summarize and describe the significant developments that characterize the three stages of the *prenatal period*. (pages 92-98)

2.   Explain how drugs, chemical agents, and environmental toxins can influence the prenatal environment.   (98-102)

3.   Briefly discuss how family and friends should respond to a couple grieving because of their miscarriage.   (100)

4.   Summarize the effects of prenatal, maternal infectious and noninfectious diseases on the fetus and on the mother.   (102-103)

5.   Briefly discuss the effects of maternal sensitization (the Rh factor). (103-104)

6.   Discuss the research regarding the effects of maternal nutrition during the *prenatal period*. (104-105)

7.   Briefly discuss the effects of maternal stress ("blood-bourne" anxieties) on the fetus. (105-106)

8.   Summarize the influence of maternal age on the unborn child.   (106)

9.   Describe the three stages of the *birth* process.   (106-108)

10. List the five conditions of the *Apgar Scoring System* and characterize each along the 0-2 rating continuum. (107-108, 110)

11. Define *natural childbirth*; identify two *natural childbirth* techniques, and discuss their advantages. (108-111)

12. Describe the relationship between birthing rooms, family-centered hospital care, and the parent-infant bonding process. (111-113)

13. Discuss some cross-cultural differences in the way fatherhood is treated. (114)

14. Briefly explain Leboyer's contemporary view of the baby's birth experience and critics' response to his claims. (113, 115)

15. Summarize the effects on mother and child of the following pregnancy and birth complications: (115-118)
    a. *toxemia*

    b. *Caesarean section*

    c. *premature infant*

16. Briefly discuss some of the developmental interventions provided for *premature infants*. (117-119)

afterbirth (page 107)

Apgar Scoring System (107)

birth (106)

blastocyst (93)

Caesarean section (115)

cephalocaudal development (94)

delivery (107)

embryonic period (92)

fetal period (92)

germinal period (92)

labor (106)

mitosis (92)

natural childbirth (108, 110)

parent-infant bonding (111)

premature infant (116)

prenatal period (92)

proximodistal development (94)

teratogen (99)

teratology (99)

toxemia (115)

SELF-TESTS   CHAPTER 4

Multiple-Choice

Circle the letter of the response which *best* completes or answers each of the following statements and questions.

Factual Questions:

1. As a result of cleavage, the zygote becomes transformed into a hollow ball of cells called

    a. mitosis
    b. the blastocyst
    c. the chorion
    d. the ectoderm

2. The placenta is the organ that

    a. transfers maternal nutrients to the fetus
    b. protects the fetus against infectious organisms
    c. directly connects the fetal and maternal blood supplies
    d. connects the embryo to the vaginal wall

3. The principle that describes development as starting with the head and brain areas and moving downward is called _____ development.

    a. cephalocaudal
    b. proximodistal
    c. anterior-posterior
    d. inverse progressive

4. During the _____ period of prenatal development, the organs of the fetus assume their specialized functions.

    a. germinal
    b. embryonic
    c. fetal
    d. critical

5. Which of the following statements concerning the effects of miscarriage is correct?

    a. The longer the pregnancy continues, the more severe the loss is.
    b. Women who miscarry may feel that they are being punished for something that happened in their past.
    c. A mother who miscarries should be told that her angry feelings are rational.
    d. Men feel less sense of loss than women do following a miscarriage.

53

6. Babies born to mothers who smoke are

   a. more likely to be mentally subnormal
   b. on the average smaller than babies of nonsmokers
   c. more likely to have neurological defects
   d. no different from babies of nonsmokers

7. Women who are exposed to gaseous anesthetics used in hospital settings are more likely than other women to

   a. have spontaneous abortions
   b. give birth to children with central nervous system disorders
   c. give birth to children with widely spaced eyes
   d. give birth to children who will later experience developmental delays

8. According to your text, which of the following is a common characteristic of mothers who give birth to an infant infected with the AIDS virus?

   a. They show outward signs of the AIDS virus.
   b. They come from a lower socioeconomic environment.
   c. They were infected through a blood transfusion.
   d. They are single.

9. Rh-negative disorder is one in which

   a. the mother develops a fever and a slight rash
   b. the mother forms antibodies that destroy the baby's blood cells
   c. the baby is born with deformed arms and legs
   d. the baby is born with facial, cardiac, and behavioral abnormalities

10. What term is used to describe the stage in the birth process in which the strong muscle fibers of the uterus contract rhythmically, pushing the infant downward toward the birth canal?

    a. delivery
    b. afterbirth
    c. labor
    d. natural childbirth

11. Crowning occurs when the

    a. widest diameter of the baby's head is at the mother's vulva
    b. Braxton Hicks contractions begin
    c. placenta is expelled from the uterus
    d. normalcy of the baby's condition at birth is checked

12. Women who use the psychoprophylactic method of childbirth

    a. use pain medication to tolerate the uterine contractions
    b. use breathing and relaxation exercises during their labor
    c. become more affectionate toward their infants
    d. have their babies delivered with sterile obstetrical equipment

13. Dissatisfaction with the maternity care options provided by physicians has contributed to the revival of

    a. birthing rooms
    b. the couvade syndrome
    c. the principle of legitimacy
    d. midwifery

14. Which of the following statements best describes family-centered hospitals?

    a. They separate the infant from the mother for a prolonged period of time following birth.
    b. They discourage new mothers from breast feeding their babies.
    c. They allow the father to be an active member of the delivery room team.
    d. They require the mother to use psychoprophylactic methods of delivery.

15. When men experience the "couvade syndrome," they are

    a. participating in a form of prepared childbirth
    b. protecting their fetus against the trauma of conventional birthing procedures
    c. experiencing pregnancy symptoms similar to those of their wives
    d. feeling a sense of loss following their wife's miscarriage

16. According to the French obstetrician Frederick Leboyer, conventional hospital birthing procedures

    a. may permanently damage the infant's nervous system
    b. produce high levels of stress hormones in the fetus
    c. increase the likelihood of delivering babies by Caesarean section
    d. ensure that the newborn infant will have an adequate Apgar rating

17. A surgical procedure by which the physician enters the uterus through an abdominal incision and removes the infant is called a

    a. uterectomy
    b. suturemia
    c. Lamaze method
    d. Caesarean section

18. The National Center for Disease Control has reported the trend of babies born with low birthweight to be

    a. decreasing slightly
    b. unchanged
    c. increasing
    d. decreasing at a dramatic rate

19. The most common difficulty encountered by premature infants is

    a. erythroblastosis fetalis
    b. respiratory distress syndrome
    c. intestinal complications
    d. intracranial bleeding

Conceptual Questions:

1. An unmarried female friend of yours thinks that she is pregnant because her period is two weeks overdue. If she is pregnant, she is probably at the

    a. end of the germinal period
    b. beginning of the fetal period
    c. middle of the embryonic period
    d. beginning of the germinal period

2. A mother experiences the phenomenon of quickening. It is also likely that at about the same time the baby's body

   a. is being covered by vernix caseosa
   b. is being covered by lanugo
   c. has rounded limbs
   d. has red, wrinkled skin

3. An unknown teratogenic substance used during pregnancy is suspected of having caused pregnant women to give birth to children with missing limbs. The likely villain would be

   a. marijuana
   b. alcohol
   c. Thalidomide
   d. vinyl chloride

4. A woman is saddened when her pediatrician tells her that her son was born with congenital deafness. She feels responsible for her son's defect because she

   a. married a man who had served in the Vietnam War and handled dioxin containers
   b. smoked heavily throughout her pregnancy even though her doctor advised her to quit smoking
   c. shared intravenous needles with a friend
   d. used quinine when serving as a missionary in equatorial Africa

5. Martha had a miscarriage the week after her obstetrician informed her that she was pregnant. The best thing her friends could tell her is

   a. "You'll get over it and look back at yourself thinking how silly you were for being so depressed."
   b. "At least you didn't miscarry after carrying the baby longer; that would have been harder to deal with."
   c. "It's OK to feel whatever it is you are feeling; go ahead and tell me how you feel because I'm willing to listen."
   d. "Now that you've had this happen to you, maybe you'll take it a little easier the next time you get pregnant."

6. Which of the following statements most accurately represents your text's discussion of toxins in the workplace?

   a. The female's ova are more susceptible to damage from environmental cause than the male's sperm.
   b. The male's sperm are more susceptible to damage from environmental toxins than the female's ova.
   c. It makes sense to clean up the workplace not only for mothers but for fathers as well.
   d. Hazards in the workplace rarely contribute to reproductive problems.

7. A pregnant woman's body produces antibodies that attack the baby's blood cells. Which of the following conclusions is most likely?

   a. The mother is more likely to be white than black.
   b. The mother is affected by toxemia.
   c. The mother is a diabetic.
   d. The mother is probably hypertensive.

8. A fetus has 20 percent fewer brain cells than normal. After birth, the same child's emotional and intellectual growth is stunted. This problem was probably caused by maternal

   a. stress
   b. Rh-negative blood
   c. diabetes
   d. malnutrition

9. Which of the following statements is *not* true regarding your text's discussion of "blood-bourne" anxieties?

   a. Women working long hours at a stressful job are more likely than other mothers to have difficulties during pregnancy.
   b. Mothers under considerable emotional stress are more likely than other mothers to experience complications during both pregnancy and labor.
   c. Prolonged emotional stress is associated with low birth-weight babies.
   d. Prolonged emotional stress is associated with infant hyperactivity.

10. We can assume that pain is a conditioned response. If we replace pain with relaxation and controlled breathing, we can

    a. increase the effectiveness of analgesic drugs
    b. increase the test performance of babies born to nonmedicated mothers
    c. decrease the need for medication during labor
    d. increase the need for analgesics during the delivery process

11. A pregnant woman is told by her friends that her baby has dropped. She also acknowledges to them that the baby feels lighter and that she has been able to breathe more easily. We can probably conclude that

    a. her baby is in the embryonic period
    b. she is having Braxton Hicks contractions
    c. crowning is about to occur
    d. her cervix is dilating

12. An anxious, expectant mother comes into the hospital maternity ward expecting to shortly deliver her second child. Her contractions are lasting ten seconds and are fifteen minutes apart. Which of the following conclusions is <u>least</u> likely?

    a. She probably has fourteen hours of hard labor ahead of her.
    b. She may expect at least eight more hours of labor.
    c. She will spend more time in labor than in delivery.
    d. She will experience stronger contractions as the contraction interval shortens.

13. Which of the following statements best exemplifies your text's discussion of parent-infant bonding?

    a. Research on early parent-child physical contact finds only minimal evidence for its lasting effects.
    b. Research suggests that parents who do not have contact with their infant immediately after delivery typically develop a weaker attachment to the youngster as do parents who do have such contact.
    c. Research on early parent-child physical contact shows significant evidence for its lasting effects.
    d. Research on mothers who have a Caesarean delivery finds significant evidence that these mothers will have difficulties in establishing a relationship with their child.

14. A laboring woman is encouraged to focus on a spot on the ceiling tile while her husband tells her to take a deep breath in, hold it, and let it out slowly. All of the following are true concerning this delivery except that

   a. natural childbirth is being used
   b. a psychoprophylactic method is being used
   c. the Lamaze technique is being used
   d. the family-centered technique is being used

15. Maria and Jose are having their first baby. A nurse-midwife will assist with the delivery, and Maria's mother will also be present for this special moment. We can also reasonably infer that

   a. Jose's active participation provides him with a sense of legitimacy
   b. Maria and the baby will be immediately separated after the birth
   c. Jose is likely to experience the "couvade syndrome"
   d. their newborn will be healthier as a result of this "birth without violence"

16. Jennifer was born in a dimly lit room. After she was born, she was immersed in tepid water and allowed to uncurl from her fetal position by being laid on her mother's bare torso. We can probably infer that, compared with other normally delivered children, Jennifer will

   a. have a higher Apgar score
   b. not be significantly different
   c. have fewer developmental delays
   d. be more closely attached to her mother

17. According to your text, the high rate of repeat Caesareans is related to which of the following?

   a. the greater profitability of the surgery
   b. the age of the physician
   c. the mother's previous painful experience with vaginal birth
   d. all of the above

18. Premature babies are often more viable and less prone to respiratory distress if they are born after the seventh month. This may be due to the presence of

   a. surfactant
   b. meconium
   c. hyaline villi
   d. bilirubin

19. A friend of yours is disappointed and concerned because she gave birth to a premature infant. She was separated from her baby for one day while he was placed in a neonatal intensive care unit. You can advise her that

   a. most low-birth-weight babies experience moderate intellectual impairment
   b. most premature babies are not growth-impaired, and she can help her baby with extra stimulation
   c. her baby will probably be physically normal, but its attachment to her will be impaired
   d. her baby will probably be physically impaired, but its attachment to her will be normal

Matching: Definitions

Match the letter of the term in the box with its definition.

---

a.	birth	g.	toxemia
b.	parent-infant bonding	h.	premature infant
c.	afterbirth	i.	teratology
d.	teratogen	j.	natural childbirth
e.	Caesarean section	k.	delivery
f.	labor	l.	Apgar scoring system

---

1. the field of study concerned with birth defects _____

2. the transition between dependent existence in the uterus and life as a separate organism _____

3. a surgical technique by which the physician enters the uterus through an abdominal incision and removes the infant _____

4. the last stage in the birth process, characterized by the expulsion of the placenta _____

5. a method for appraising the normalcy of a baby's condition at birth _____

6. any agent that contributes to birth defects _____

7. stresses the active involvement of the mother and father in the childbirth process _____

8. a baby weighing less than 5 1/2 pounds at birth or having a gestational age of less than 37 weeks _____

9. a disorder of pregnancy characterized by high blood pressure (hypertension) _____

10. the notion that closeness between parent and child in the first minutes of life produces a lasting effect _____

11. a stage in the birth process in which the baby is pushed down the birth canal _____

12. a stage in the birth process that begins once the infant's head passes through the cervix _____

Matching: Stages of Prenatal Development

Match the letter of the term in the box with its definition.

a.	embryonic period	e.	blastocyst
b.	mitosis	f.	germinal period
c.	cephalocaudal development	g.	proximodistal development
d.	prenatal period	h.	fetal period

1. phase that extends from about the end of the second week following conception to the end of the eighth week _____

2. development away from the central axis of the organism _____

3. the time elapsing between conception and birth _____

4. the phase that extends from the end of the eighth week until birth _____

5. a hollow ball of cells that develops from a fertilized ovum _____

6. the phase that extends from conception to the end of the second week _____

7. development starting with the brain and head and then working its way down the body _____

8. the process during which the zygote divides and forms two cells identical in makeup with the first cell _____

Completion

Supply the term needed to complete each of the following statements.

1. During the germinal period of prenatal development, the blastocyst separates into an outer layer of cells called the _____ and an embryonic inner cell mass.

2. The spontaneous movements of the fetus are referred to as _____.

3. In their _____, fathers in some societies seclude themselves during and after the period of childbirth.

4. Prematurely born infants who suffer from respiratory distress syndrome lack a lung lubricant known as _____.

5. The transfer surface between the placenta and embryo has _____, a web of fingerlike projections that extend into blood spaces in the uterus.

6. _____ is another word for downy, fetal hair.

7. The _____ is a hollow ball of cells.

8. The point in pregnancy when the uterus sinks down and forward is called _____.

9. The early cell divisions in development are called _____.

10. The _____ is the term used to describe the wall of the uterus.

Multiple-Choice

Factual

1.	b	(93)
2.	a	(93-94)
3.	a	(94)
4.	c	(95)
5.	b	(100)
6.	b	(99)
7.	a	(101-102)
8.	b	(103)
9.	b	(104)
10.	c	(106)
11.	a	(107)
12.	b	(110)
13.	d	(113)
14.	c	(113-114)
15.	c	(114)
16.	a	(115)
17.	d	(115)
18.	c	(116)
19.	b	(116)

Conceptual

1.	c	(95)
2.	b	(95)
3.	c	(99)
4.	d	(99)
5.	c	(100)
6.	c	(101-102)
7.	a	(104)
8.	d	(104)
9.	a	(105)
10.	c	(111)
11.	b	(106)
12.	a	(106-107)
13.	a	(112)
14.	d	(112-114)
15.	a	(114)
16.	b	(115)
17.	d	(115-116)
18.	a	(117)
19.	b	(118-119)

Matching: Definitions

1.	i	(99)
2.	a	(106)
3.	e	(115)
4.	c	(107)
5.	l	(107-108)
6.	d	(99)

7.	j	(108, 110)
8.	h	(116)
9.	g	(115)
10.	b	(111)
11.	f	(106)
12.	k	(107)

Matching: Stages of Prenatal Development

1.	a	(92)
2.	g	(94)
3.	d	(92)
4.	h	(92)

5.	e	(93)
6.	f	(92)
7.	c	(94)
8.	b	(92)

Completion

1.	trophoblast (93)
2.	quickening (95)
3.	couvade (114)
4.	surfactant (117)
5.	villi (94)

6.	lanugo (95)
7.	blastocyst (93)
8.	lightening (106)
9.	cleavage (93)
10.	endometrium (93)

Chapter 5

The Newborn

INTRODUCTION

Chapter 4 examined the period of prenatal development and the birth process. Chapter 5 highlights the abilities and behavior of neonates as expressed during the first two weeks following birth. Several important topics are discussed, including:

- Newborn Behaviors and Abilities. This section includes an in-depth discussion of newborn behaviors and abilities, including sleeping, crying, feeding, reflexes, memory, and learning.

- Senses. An overview is presented of the infant's ability to see, hear, taste, smell, and feel heat, cold, pressure, and pain. Research related to the significance of these senses is cited.

- Individual Differences. Newborns are individually different from one another, and the research is examined that leads to this conclusion.

I. Newborn Behaviors

A. Sleeping

B. Crying

C. Feeding

D. Reflexes

E. Memory: Habituation

F. Learning

II. The Senses

A. Vision

B. Hearing

C. Taste and Smell

D. Cutaneous Senses — *of or relating to the skin*

III. Individual Differences Among Newborns

A. Cuddlers and Noncuddlers

B. Differences in Temperament

C. Parent-Child Interaction: A Two-Way Street

LEARNING OBJECTIVES

After completing Chapter 5, you should be able to:

1. Briefly describe the *neonate's* appearance. (page 124)

2. Discuss what is known about Sudden Infant Death Syndrome (SIDS). (Box 125)

3. Identify and describe the six types of *neonatal* states illustrated by Wolff. (pp. 125-126)

   a.

   b.

   c.

   d.

   e.

   f.

4. Recognize the significance of the infant's crying and the effect of the caregiver's response on the newborn's crying. (127-130)

5. List a variety of techniques that parents have acquired to help soothe their infants. (127)

6. Describe the effects of drug use by pregnant women on the newborn. (Box 128-129)

7. Summarize what is known about the advantages and disadvantages of schedule versus demand feeding and breast versus bottle feeding. (130-131)

8. Explain the significance of the following *reflexes*: (131-133)

   a. Moro

   b. startle

   c. rooting

   d. sucking

64

e. Babinski

f. Palmar

g. plantar

h. stepping

i. swimming

9. Describe the importance of *habituation* in the *neonate* and provide an example to illustrate this behavior. (133-34)

10. Discuss the significance of learning in the newborn and given an example of this behavior. (134-35)

11. Define *sensation* and *perception*. (135)

12. Establish a contemporary view of the *neonate's* sensory capabilities by summarizing the research of Fantz, Bower, Meltzoff, Moore, and Jacobson on infant visual abilities. (136)

13. Evaluate the *neonate's* ability to see, hear, taste, smell, and feel heat, cold, pressure, and pain. (135-139)

14. Discuss what is known about circumcision. (Box 140)

15. Categorize the individual differences observed in newborns regarding cuddling and *temperament*, and discuss their significance. (139-142)

16. Summarize Alexander Thomas's "goodness of fit" concept. (141-142)

17. Discuss the two-way interaction that characterizes parent-child relationships. (142-144)

KEY TERMS

---

habituation (page 134)

neonate (124)

neonate period (124)

perception (135)

reflex (131)

sensation (135)

state (124)

temperament (140)

Multiple-Choice

Circle the letter of the response which *best* completes or answers each of the following statements and questions.

Factual Questions:

1. The heads of many newborns are misshapen and elongated as a consequence of passage through the birth canal. We call this condition

    a. shaping
    b. quickening
    c. states
    d. molding

2. Which of the following has <u>not</u> been suggested as a possible cause of Sudden Infant Death Syndrome (SIDS)?

    a. breast feeding
    b. apnea
    c. bacterial infection
    d. brain defects

3. Because they lack strong clues that would identify babies in danger of developing SIDS, researchers have recently turned to theories of

    a. oxygen deprivation
    b. prenatal care
    c. postnatal care
    d. brain development

4. During the state that Peter Wolff refers to as "drowsiness," the infant

    a. is at full rest
    b. has occasional and rapid eye movement
    c. is relatively inactive but opens and closes its eyes intermittently
    d. has irregular respiration and frequent, diffuse motor activity

5. Which of the following descriptions would *best* characterize an infant's crying at the age of 9 months?

    a. fussy and irregular
    b. persistent and frequent
    c. punctuated by pauses
    d. rhythmic

6. Researchers have discovered that babies who cry little and whose crying decreases during the year generally have mothers who

    a. ignore them when they cry
    b. respond to them when they cry
    c. distract the infant with novel visual stimulation
    d. have a high tolerance for frustration

7. Which of the following statements regarding the effects of drug use by pregnant women on their newborn infants is *not* true?

   a. Since most states deem a fetus to be a child, child abuse statutes are now being applied to fetal abuse.
   b. Penalties issued by the courts tend to fall more heavily on poor minority women using drugs.
   c. Most schools are not set up to provide the kind of atmosphere and nurturing needed by drug-damaged children.
   d. The March of Dimes reports that the number of crack-exposed children could reach as high as 4 million by the year 2000.

8. One of the major advantages of breast versus bottle feeding is that breast feeding

   a. gives mothers more physical freedom
   b. gives babies better startle reflexes
   c. immunizes babies by medication the mother receives
   d. stimulates maturation of the infant's intestinal tract

9. Reflex responses in infants

   a. are primitive mechanisms that serve no useful biological function
   b. are generally necessary for human survival
   c. provide the infant's limbs with needed exercise
   d. are good indicators of neurological development

10. The rooting reflex in infants is activated by touching the

   a. palm of the infant's hand
   b. corner of the infant's mouth
   c. soft palate of the mouth
   d. outer edge of the sole of the foot

11. In order to demonstrate the phenomenon of habituation, an infant must

   a. stop responding to a repeated stimulus and then recover its response
   b. learn to change its behavior in order to produce new consequences
   c. accommodate the lens of its eye meaningfully
   d. move its head defensively in response to a moving object

12. Arnold Sameroff conducted a study in which he was able to teach neonates to use one method of sucking (either expressive or suction) in preference to another. The learning was achieved through

   a. modeling
   b. observation
   c. punishment
   d. reinforcement

13. Which of the following stimuli is *least* likely to attract the visual attention of an infant?

   a. facelike pattern
   b. slow-moving object
   c. plain shape
   d. black-and-white bull's-eye pattern

14. A psychologist projects a virtual visual image in front of 2-week-old infants. The infants will probably

   a. not detect the difference between the virtual image and the real image
   b. open their mouths and stick out their tongues
   c. react with surprise when their hands reach the location where the object seems to be
   d. turn their heads and focus on another more interesting three-dimensional stimulus

15. William Condon and Louis Sander have videotaped neonate-adult interaction. They find that neonates

   a. do *not* possess a genetic predisposition for the acquisition of language
   b. synchronize (coordinate) the movement of their bodies to adult speech patterns
   c. hold their breath and tense their bodies when adults start to speak
   d. have an incompletely developed hearing apparatus resulting from vernix in the ear

16. When a sweet fluid is put into the mouth of newborns, they

   a. decrease sucking speed
   b. use the expression sucking method
   c. display increased respiration rate
   d. engage in increased limb movements

17. H. R. Schaffer and Peggy Emerson studied the reactions of 37 neonates to physical contact. They found that

   a. noncuddlers are typically ahead of cuddlers in motor development
   b. noncuddlers object to being swung, bounced, or romped with
   c. cuddlers crawl sooner than noncuddlers
   d. cuddlers object to close physical contact

18. According to Alexander Thomas's "goodness of fit" concept, a good match is one in which the opportunities, expectations, and demands of the environment are in accord with the child's

   a. self-concept
   b. temperament
   c. developmental level
   d. activity level

19. Researchers have found that mothers of firstborn infants spend more time
   _____ their infants than mothers of later-born children.

   a. holding
   b. talking to
   c. feeding
   d. singing to

Conceptual Questions:

1. Michelle and Russell are grieving the loss of their 3-week-old son. After the autopsy was performed, the doctor told this couple that the respiratory control center of their infant's brain may have been defective. We might also reasonably infer that their son experienced

   a. suffocation
   b. deficient suction
   c. apnea
   d. poor prenatal care

2. Juan and Maria observe that their newborn baby girl's face is contorted, and she is making strong and intense vocalizations while moving about vigorously. According to Wolff, the infant is in which *state*?

   a. irregular sleep
   b. alert inactivity
   c. waking activity
   d. crying

3. Your friend Mary is concerned because her baby cries frequently and makes her edgy. Mary is afraid that if she picks her baby up all the time, the baby will become spoiled. Your best advice to her would be that

   a. temperamental babies should be handled firmly and ignored
   b. crying is good for children because it helps them expand their lungs
   c. her baby's crying is a temperamental problem, and there is really nothing she can do about it
   d. simply picking her baby up to her shoulder will pacify the baby without spoiling it

4. Which of the following statements is most consistent with the issue of how to deal with mothers who damage their children by taking drugs?

   a. Since most states deem a fetus to be a child, child abuse statutes apply to fetal abuse.
   b. The mothers should be punished by our nation's courts.
   c. The best solution is prevention and early treatment.
   d. Children have a constitutional right to be born drug-free.

5. Select the group of terms that can be most logically placed in the same category:

   a. self-demand feeding, longer feeding intervals, vitamin D
   b. breast feeding, antibody production, intestinal tract maturation
   c. schedule feeding, bottle feeding, fluoride supplements
   d. breast feeding, malnutrition, intimate contact

6. Monique decides that her infant should be breast-fed rather than bottle-fed. Monique should *not* expect which of the following in regard to the advantages of breast feeding as opposed to bottle feeding?

   a. Breast-fed infants are easier to console than bottle-fed youngsters.
   b. Breast feeding provides emotional and psychological rewards that are not available to women who bottle feed their infants.
   c. The physical act of nursing is a sensually pleasurable experience.
   d. Breast-fed neonates have a more optimal physiological organization than bottle-fed infants.

7. A nursing mother tries to get her child to suckle on the breast by pushing the child's head toward the breast. The child attempts to suck the mother's hand, thus indicating

   a. the orienting reflex
   b. dishabituation
   c. the rooting reflex
   d. breast sensitization

8. Babinski is to _____ as plantar is to foot pressure.

   a. mouth touch
   b. palm pressure
   c. neck twist
   d. foot stroking

9. A visual display of two dots is repeatedly presented to an infant until the infant stops responding. When a new stimulus containing three dots is exposed, the infant starts responding again, indicating

    a. adaptation
    b. habitual reflexes
    c. habituation
    d. conditioning

10. A child stops orienting to a buzzer that he has repeatedly heard. The cessation of the orienting response indicates all of the following *except*

    a. high risk of mental retardation
    b. early adaptive capabilities
    c. development of early memory processes
    d. central nervous system reflex inhibition

11. Immediately after an infant kicks its right foot in the air, an experimenter activates a turning mobile above the child's head. Subsequently, the child increases the frequency of right-leg kicks but not left-leg kicks. The experimenter has demonstrated

    a. habituation
    b. sensory perception
    c. learning
    d. sensitization

12. You are designing a clown costume for Halloween. If you want to maximize an infant's visual attention, you should

    a. use very dim pastel colors in the clown's mask
    b. paint two dark-blue spots on the cheeks of the clown mask
    c. paint two straight lines on each cheek of the clown mask
    d. paint one brightly colored bull's-eye in the middle of the clown's mask

13. A young infant reaches out to touch a soap bubble and succeeds in bursting the bubble. We can reasonably predict that the infant will

    a. change its facial expression and act surprised
    b. be unresponsive to this event because she lacks visual accommodation
    c. search for the missing bubble
    d. display rapid eye movements and begin to cry

14. The best song title to describe how infants respond to the sound patterns of adult speech would be

    a. "Do You Wanna Dance?"
    b. "Hungry Eyes"
    c. "Stand by Me"
    d. "Everything I Do, I Do for You"

15. Your oldest son liked being hugged and snuggled. However, your newborn infant squirms and kicks when you cuddle her. On the basis of psychological research, you would conclude that

    a. maybe your newborn infant would prefer being bounced and tickled
    b. if you are patient, in a few weeks your newborn will learn to prefer being cuddled
    c. if your newborn infant had another type of mother, she would like being cuddled
    d. your newborn will develop more slowly than your oldest son did

16. Your infant nephew spits out new foods, has violent tantrums, and is difficult to pacify. On the basis of current research, one likely conclusion is that

    a. his parents will be likely to perceive the baby as needing greater amounts of love and attention
    b. he will be as cheerful as an easy child in another year or two
    c. his parents will likely feel considerable anxiety and guilt
    d. his behavior is primarily due to his parents' domineering, authoritarian behavior

17. A new mother is ecstatic because her infant daughter has a cheery disposition and adapts well to new situations. On the basis of recent psychological studies, she can conclude that

    a. her excellent parenting skills and positive attitudes are responsible for her daughter's disposition
    b. any other child of hers would probably react in the same manner as her daughter does
    c. her pattern of interaction with her daughter is partially a cause and partially a result of her daughter's constitutional characteristics
    d. her pattern of interaction with her child is influenced primarily by the fact that her child is a girl

Matching: Definitions

Match the letter of the term in the box with its definition.

a.	Babinski	j.	perception
b.	apnea	k.	neonate
c.	Moro	l.	state
d.	regular sleep	m.	habituation
e.	sensation	n.	neonate period
f.	reflex	o.	noncuddlers
g.	temperament	p.	palmar grasp
h.	molding	q.	olfactory
i.	virtual image	r.	drowsiness

1. arms thrown out and brought together again _____

2. sense of smell _____

3. advanced motor development _____

4. at full rest, with little or no motor activity _____

5. head distortion necessary to accommodate birth _____

6. response decrement and recovery associated with stimulus presentation and removal _____

7. reception of information by sense organs _____

8. an optical illusion _____

9. interpretation of sensory information _____

10. a newborn _____

11. pauses in breathing _____

12. two weeks following birth _____

13. continuum of alertness (sleep-vigorous activity) _____

14. simple, involuntary, unlearned response _____

15. triggered by touching the palm of the infant's hand _____

16. inherent dispositions that modulate an infant's behavior _____

17. relatively inactive, with eyes opening and closing intermittently _____

18. triggered by gently stroking the outer edge of the sole of the foot _____

Completion

Supply the term or terms needed to complete each of the following statements.

1. When a neonate is in a state of _____, she is inactive and her eyes are open and have a bright, shining quality.

2. Bell and Ainsworth discovered that ignoring babies' cries actually _____ the likelihood that they will cry more as the first year progresses.

3. The use of _____ feeding has been linked with malnutrition in third-world infants.

4. The _____ reflex is activated when the neonate hears a loud noise or is suddenly touched.

5. By displaying habituation, infants give us insight into their _____ processes.

6. An infant's typical response when he is confronted with a virtual visual image may indicate that some aspects of _____ coordination may be biologically prewired.

7. When infants first detect a(n) _____, they move their limbs, they breathe quickly, and their heart rate increases.

8. When we refer to the _____ of _____, we are describing the match between an infant's temperamental characteristics and the opportunities, expectations, and demands of its environment.

9. Psychologists have observed that parenting is easier for predictable and cheery babies, thus suggesting that the socialization process is _____.

10. Infants who die of Sudden Infant Death Syndrome (SIDS) may be suffering from _____.

Multiple-Choice

Factual

1.	d	(124)
2.	a	(125)
3.	d	(125)
4.	c	(126)
5.	c	(129)
6.	b	(127)
7.	a	(128-129)
8.	d	(130)
9.	d	(131)
10.	b	(132)
11.	a	(133-134)
12.	d	(134)
13.	c	(136)
14.	c	(136)
15.	b	(137-138)
16.	a	(138)
17.	a	(140)
18.	b	(142)
19.	c	(144)

Conceptual

1.	c	(125)
2.	d	(126)
3.	d	(127)
4.	c	(128-129)
5.	b	(130)
6.	a	(130-131)
7.	c	(132)
8.	d	(132)
9.	c	(133-134)
10.	a	(134)
11.	c	(134)
12.	d	(136)
13.	a	(136)
14.	a	(138)
15.	a	(140)
16.	c	(142)
17.	c	(142-143)

Matching: Definitions

1.	c	(131-132)		10.	k	(124)
2.	q	(138)		11.	b	(125)
3.	o	(140)		12.	n	(124)
4.	d	(125)		13.	l	(124)
5.	h	(124)		14.	f	(131)
6.	m	(133-134)		15.	p	(132)
7.	e	(135)		16.	g	(140)
8.	i	(136)		17.	r	(126)
9.	j	(135)		18.	a	(132)

Completion

1.	alert inactivity (126)		6.	eye-hand (136)
2.	increases (127)		7.	odor (138)
3.	bottle (131)		8.	goodness; fit (141-142)
4.	startle (132)		9.	reciprocal or interactive (143-144)
5.	memory (134)		10.	apnea (125)

Chapter 6

Infancy: Development of Basic Competencies

INTRODUCTION

Chapter 6 explores the period of infancy--the first two years of life--during which children interact with their environment to develop basic competencies. As children age, they are continuously forced to adapt to the demands of the world around them. Chapter 6 focuses on the three areas of development which are most profound during the first two years of life: physical growth, motor development, and perceptual development.

- Physical Growth. This generally occurs in an orderly fashion, but factors such as malnutrition and disease can interfere with proper development. Not all parts of a child's body develop at equal rates. The brain grows rapidly during the first two years of life. Development in all children is known to follow the cephalocaudal principle as well as the proximodistal principle.

- Motor Development. This is highly complex in infants and is dependent upon the child's overall physical growth. Rhythmical behaviors, such as kicking and rocking, provide the basis for more skilled motor developments, such as crawling and walking. The timing of motor development differs widely among cultures. Handedness is a motor skill which manifests itself by the end of infancy. The rarer left-handed trait may derive primarily from prenatal, genetic, or learning factors.

- Perceptual Development. This peaks during the first six months of life, prior to the large spurt in motor development witnessed in the seventh month. Visual constancy is apparent in children during these first months of life as documented by Gibson's visual cliff experiment. Most babies possess depth perception by the time they reach one year of age. How infants focus on and organize visual events changes dramatically during the first two years of life. Children under one year of age prefer novel stimuli, whereas older children focus more on very unusual stimuli. Not all patterns of development occur in a strictly increasing fashion, as in eye-hand coordination which follows noncumulative patterns of skill acquisition, loss, and reacquisition.

I. Physical Growth

    A. Rate of Growth

    B. Malnutrition

    C. Secular Increase in Size

    D. Growth of Key Systems and the Brain

    E. Principles of Development

II. Motor Development

    A. Rhythmical Behaviors

    B. Locomotion

    C. Manual Skills

    D. Handedness

III. Perceptual Development

    A. Visual Constancy

    B. Depth Perception

    C. Perception of Form

    D. Perception of the Human Face

    E. Interconnections Among the Senses

LEARNING OBJECTIVES

After completing Chapter 6, you should be able to:

1. Define the following terms; explain their significance regarding physical growth: (page 148)
    a. *infancy*

    b. pituitary gland

    c. hypothalamus

    d. *norms*

    e. sequence

    f. growth spurts

    g. target seeking

2. List the characteristics and consequences of the following disorders: (149-150)

    a. marasmus

    b. kwashiorkor

3. Explain how early inadequate nutrition affects development. (150)

4. Evaluate the data with regard to whether children will instinctively select a well-balanced diet. (Box 152-153)

5. Discuss the status of future generations with regard to current living conditions in the United States for young children. Consider such factors as infant mortality rates, disease, and immunization. (Box 154-155)

6. Describe the *secular increase in size* and the two theories developed to explain it. (150-151)

7. Briefly discuss how distinct parts of the body grow at different rates. (151-152)

8. Identify the significance of the following terms with regard to brain development: (152-153)

    a. cerebral cortex

    b. subcortical level

    c. left hemisphere

    d. right hemisphere

9. Briefly explain the changes in brain activity provided by the positron emission tomography (PET) procedure. (157-158)

10. Describe the *cephalocaudal* and *proximodistal principles* of development, and provide examples of each with regard to motor behavior. (157-158)

11. List some of the rhythmical behaviors displayed by young infants, and discuss their role in motor development. (159-160)

12. Trace the development of locomotion, specifically emphasizing the role of the following: (160-161)

    a. crawling

    b. creeping

    c. hitching

    d. cultural factors

13. Briefly describe the child's development of manual skills. (162-163)

14. Describe the development of handedness, and discuss why left-handers suffer greater risks than right-handers. (163-164)

15. State how testosterone affects the development of brain structures. (164)

16. Define *visual constancy*, and summarize the research evidence on its development throughout infancy. (164-165)

17. Explain Gibson's *visual cliff experiment*, and state its significance with respect to depth perception in infants. (166)

18. Describe the following terms as they apply to perception of form during infancy: (167-168)

   a. visual scanning

   b. corneal photography

   c. novel stimuli

   d. unusual stimuli

   e. hypothesis

19. Summarize the infant's changing perception of the human face that occurs during the first year of life. (168-169)

20. Explain how eye-hand coordination may develop in light of the theories proposed by Piaget and Brower. (169-171)

KEY TERMS

---

cephalocaudal principle (page 157)

infancy (148)

norms (148-149)

proximodistal principle (158)

secular increase in size (150-151)

visual cliff experiment (166)

visual constancy (165)

Multiple-Choice

Circle the letter of the response which *best* completes or answers each of the following statements and questions.

Factual Questions:

1. The two structures known to play a critical role in the growth regulation of children are

    a. the pituitary gland and the hypothalamus
    b. the pituitary gland and the cerebral cortex
    c. the subcortical level and the hypothalamus
    d. the left hemisphere and the right hemisphere

2. A child who suffered an acute illness as an infant recovers and resumes his normal growth curve. This recovery is possible because growth is

    a. norm-seeking
    b. controlled by the pituitary
    c. self-stabilizing
    d. maturationally sequential

3. Children who develop marasmus or kwashiorkor

    a. have diets that are deficient in calcium
    b. never entirely catch up with healthy children in stature or intellectual performance
    c. will recover with injections of human growth hormone
    d. will develop a protuberant belly resulting from an excessive intake of high-caloric food

4. Which of the following is affected by inadequate early nutrition?

    a. brain weight
    b. intelligence
    c. number of brain cells
    d. all of the above

5. Which of the following statements regarding nutrition in young children is the *least* accurate?

    a. Rewarding a child for eating a specific food tends to increase the child's preference for it after the reward stops.
    b. Youngsters who do not eat as much at one meal will make up for it at another meal.
    c. Dietary patterns established early in childhood have consequences for the later development of various diseases.
    d. They have the capacity to learn what is not food and what should not be ingested.

6. Which of the following statements concerning childhood illness is correct?

    a. AIDS is not considered a major cause of death among small children worldwide.
    b. Ninety percent (90%) of the world's children receive vaccinations against the most common childhood diseases.
    c. Entering the 1990s, the United States had, by far, the lowest infant mortality rate in comparison with other developed nations.
    d. Respiratory infections are currently the leading cause of death among infants and small children worldwide.

7. The secular increase in size phenomenon is

   a. not attributable to environmental factors such as diet and health
   b. showing signs of leveling off in countries such as the U.S.
   c. not witnessed in countries other than the U.S.
   d. seen predominantly in males, while the average female height has remained constant

8. The lymphoid tissue shows its greatest development _____, whereas growth of the reproductive system peaks _____.

   a. prior to adolescence; during adolescence
   b. during adulthood; during adolescence
   c. prior to adolescence; during adulthood
   d. during adulthood; during early infancy

9. Neurological research indicates that

   a. the corpus callosum in the newborn prevents passage of information between the two cerebral hemispheres
   b. rapid cortical development in infancy allows infants to develop more flexible and less stereotyped behaviors
   c. the right hemisphere is typically specialized for reasoning operations and speech control
   d. most reflexes, such as sucking and grasping, are organized in the cerebral cortex of the newborn's brain

10. According to the cephalocaudal principle, development proceeds

    a. from near to far
    b. from feet to head
    c. from head to feet
    d. from far to near

11. Recent research on motor development in infants indicates that

    a. behaviors such as crawling emerge from the dynamic interplay of several developing capabilities
    b. capacities such as locomotion typically emerge in a linear manner
    c. the capacity for locomotion is independent of the child's overall physical growth
    d. the capacity for locomotion is unaffected by the child's interaction with the world around him

12. In learning to walk, children progress in a long sequence of developments. These developments

    a. progress in a sequence which follows the cephalocaudal principle
    b. are initiated by the infant's command of the trunk region
    c. surface most noticeably during the fourth month of life
    d. progress in sequence from hitching, to creeping, to crawling during the seventh month

13. Recent studies of African infants indicate that

    a. the timing of motor development is very similar among all cultures
    b. African babies are slow in motor development relative to Caucasian infants
    c. the African infants' accelerated motor development is attributed to environmental factors such as stimulating child-care practices
    d. genetic explanations alone account for the differences in motor development timing among cultures

14. Which of the following statements concerning handedness is accurate?

    a. Handedness develops gradually, with children under 2 years of age displaying versatility in using both hands.
    b. By the time they are 2 years of age, most children shift toward left-handedness.
    c. Ambidexterity is relatively uncommon in children under 4 years of age.
    d. Human beings are decidedly biased toward the right in their lateral preferences.

15. Left-handers suffer elevated risk in comparison with their right-handed counterparts because of several factors. Which of the following does *not* play a role in this susceptibility?

    a. elevated accident risk
    b. testosterone levels
    c. immune disorders
    d. estrogen levels

16. Jerome Kagan finds in his research that infants between 3 and 12 months of age are attracted most of all to stimuli that are

    a. moderately novel and can be assimilated to their schemes
    b. of the visual cliff type
    c. very unfamiliar and repetitive
    d. very familiar but not encountered frequently

17. Jerome Kagan suggests that as children approach one year of age, they develop a new approach to stimuli, involving

    a. the use of a scheme to attend to or ignore a stimulus
    b. periods of concentrated gazing alternated with periods of avoidance
    c. the formulation of hypotheses regarding the nature of incoming visual stimulation
    d. periods of concentrated gazing accompanied by elevated heart rates

Conceptual Questions:

1. Your child appears very healthy, but he is much smaller than most children his age. The best advice might be to have your child's

    a. target-seeking rate monitored
    b. pituitary gland and hypothalamus checked
    c. testosterone levels checked
    d. doctor prescribe large doses of growth hormone to stimulate growth

2. Which of the following situations might cause the greatest amount of concern to parents?

    a. Their child has grown 20 inches between ages 4 and 5 but only grew 10 inches between ages 1 and 2.
    b. Their child has grown only 3 inches each year between ages 5 and 11.
    c. Their 12-year-old son is shorter than their 11-year-old daughter.
    d. Their child grew 4 inches between ages 4 and 5 but only 3 inches between 5 and 6.

3. Mika is a Cambodian child who has suffered from severe malnutrition. With proper food and medical care, Mika

    a. will undergo an altered physical growth pattern
    b. may demonstrate a target-seeking quality in her growth but will never entirely catch up with healthy children
    c. will experience an accelerated adolescent growth spurt completely compensating for her prior malnutrition
    d. will achieve an intellectual level equal to that of children who never suffered malnutrition

4. Ramone, a 4-year-old, refuses to eat a variety of nutritional foods including milk, eggs, and fruit. His parents should

   a. tell Ramone that he will be rewarded for eating such foods
   b. continue to let Ramone select his own food choices, regardless of their nutritional content
   c. take away all sweets from Ramone's diet until he agrees to eat healthy foods as well
   d. limit the amount of sweets in his diet and continue to provide basic foods, realizing that many toddlers experience erratic eating behaviors

5. Your father was taller than your grandfather, and you are taller than your father. Which of the following conclusions would be least valid?

   a. The generational increase in family stature may be due to improved diet.
   b. The generational increase in family stature may be due to improved prenatal care.
   c. Genetic factors play no role in this trend toward increased size.
   d. The general increase in family stature may be due to more rapid central nervous system maturation.

6. Your 2-year-old child has recently undergone PET (positron emission tomography) scanning, which maps the brain's metabolic activity. The results will probably indicate

   a. the metabolic rate of your 2-year-old's brain is about two-thirds that of an adult's brain
   b. the metabolic rate of your child's brain will never exceed that of an adult's brain
   c. the metabolic rate of your 2-year-old's brain approximates that of an adult's, because of rapid increase in cerebral cortex activity
   d. the metabolic rate of your 2-year-old's brain approximates that of an adult's, because of gradual increases in activity of the subcortical level of the brain

7. When teaching your 6-year-old son to swing a bat at a baseball, you notice that he can coordinate the movements of his forearms, but his wrist snapping is sloppy, and he holds the bat loosely. Your son is demonstrating the

   a. cephalocaudal principle
   b. effects of handedness
   c. proximodistal principle
   d. rhythmical behavior principle

8. Rhythmical behavior is to _____ as palm grasp is to _____.

   a. crawling; forefinger grasp
   b. uncoordinated activity; voluntary motor control
   c. locomotion; self-stimulation
   d. locomotion; manual dexterity

9. You adopt an Indian baby and are concerned because the baby began walking later than your biological children. A likely explanation for this difference is that

   a. Indian children begin walking later than Anglo children because such exploratory behavior is often dangerous for youngsters raised in the Indian culture
   b. your adopted Indian child was born prematurely
   c. as a newborn, the Indian child may not have been breast fed, as were your biological children
   d. Indian children begin walking later than Anglo children because Indian newborns receive a lack of maternal attention

10. Stacie begins to exhibit a strong preference for her left hand by the time she is 4 years old. Most likely, Stacie

   a. will excel in math and related areas
   b. was exposed to low levels of testosterone
   c. will excel in verbal skills
   d. will live ten years longer than her right-handed sister

11. Your son was exposed to high levels of testosterone as a fetus. What effects might this exposure have on his future?

   a. He will develop into a right-hander.
   b. He will develop a stronger immune system than those children exposed to low levels of testosterone during prenatal development.
   c. His chances of developing a learning disorder will be decreased.
   d. He will die earlier than those children exposed to low levels of testosterone during prenatal development.

12. An 8-month-old infant creeps over to the top of the staircase. If the infant avoids creeping over the edge of the top step,

   a. visual constancy has been demonstrated
   b. binocular vision has not yet occurred
   c. tactile information has been relied on more than visual information
   d. the visual cliff effect has been demonstrated

13. A 4-month-old baby looks at a photograph of a face missing a nose and does not seem surprised, whereas a 12-month-old infant looking at the same picture raises his eyes and eyebrows and shows concentrated attention. This reaction shows that the

   a. 4-month-old has sensation but not perception
   b. 12-month-old perceives a face instead of just a combination of sensory features
   c. 12-month-old is inventing hypotheses to explain a discrepant event (e.g., missing nose)
   d. 4-month-old activates hypotheses but has no standard to compare them to

14. A 4-month-old baby is shown two photographs: one of his mother's face and one of a stranger's face. We can reasonably infer that

   a. the infant cannot yet discriminate between familiar and unfamiliar faces
   b. the infant can recognize his mother from somebody else
   c. the baby will inspect only the external features of the faces, such as the chin and ears
   d. the baby will not concentrate on the pictures because of his lack of visual preference for the human face

15. You observe that your infant was able to grasp an object in the dark when he was 5 months old but fails to do so at 7 months. You should not be alarmed because

   a. older infants rely on visual feedback to enhance their reaching ability
   b. the grasping of older infants is reflexive, while that of younger infants is visually guided
   c. intersensory coordination is usually not evident in 9-month-olds
   d. 9-month-olds rely primarily on auditory motor coordination

Matching: Definitions

Match the letter of the term in the box with its definition.

```
a.  norms                       i.  secular increase in size
b.  proximodistal principle     j.  visual constancy
c.  visual cliff experiment     k.  cephalocaudal principle
d.  marasmus                    l.  motor development
e.  infancy                     m.  subcortical level
f.  pituitary gland             n.  testosterone
g.  respiratory infections      o.  binocular vision
h.  cerebral cortex             p.  perceptual development
```

1. the first two years of life _____

2. develops vastly during the first six months of life _____

3. an increase in the average stature of children _____

4. the part of the brain responsible for learning, thinking, reading, and problem solving _____

5. undergoes a big spurt around the seventh month of life _____

6. standards for evaluating a child's progress relative to the average of the child's age group _____

7. development proceeds from the head to the feet _____

8. the ability to tell the distances of various objects and to experience the world three-dimensionally _____

9. the tendency for objects to look the same to us despite fluctuations in sensory input _____

10. suggests that infants can perceive depth _____

11. the leading cause of death among infants and small children worldwide _____

12. high levels may slow growth on the left side of the brain during the prenatal period _____

13. the part of the brain that guides basic biological functioning, including sleeping, heart rate, hunger, and digestion _____

14. plays a critical role in regulating children's growth _____

15. develops as a result of protein and calorie malnutrition _____

16. development proceeds from near to far _____

Completion

Supply the term or terms needed to complete each of the following statements.

1. A child who was malnourished as an infant begins to gain weight and height rapidly after he is placed on a nutritious diet. His ability to make up for his arrested growth indicates that growth is _____ - _____.

2. UNICEF estimates that most of the world's children could be saved from death and malnutrition at a cost of _____ a year.

3. Most reflexes are organized at the _____ level of the brain.

4. The corpus callosum carries messages between the _____ and the _____.

5. According to the _____ principle of development, a child should gain control of her arms before she gains control of her legs.

6. When children sit and slide along the floor by "digging in" and pushing themselves backward with their heels, they are _____.

7. High levels of testosterone during the prenatal period may lead to an increased risk for _____ disease, as well as an increased _____ ability.

8. When shown a cross inside of a circle, infants younger than 5 months of age are more likely to see it as a cross _____ a circle, not as a cross _____ a circle.

9. In discussing children's form perception, Kagan has used the term _____ to refer to a child's mentally transforming a visual event into an already familiar form.

10. Piaget's theory that infants gradually achieve an integration of eye-hand activities only as they interact with their environment is in stark contrast to the viewpoint that eye-hand coordination is _____ in the infant's nervous system at birth.

Multiple-Choice

Factual

1.	a	(148)
2.	c	(149)
3.	b	(149-150)
4.	d	(150)
5.	a	(153)
6.	d	(154)
7.	b	(150)
8.	a	(151-152)
9.	b	(153)
10.	c	(157)
11.	a	(158-159)
12.	a	(157-159)
13.	c	(160)
14.	a	(163-164)
15.	d	(164)
16.	a	(168)
17.	c	(168)

Conceptual

1.	b	(148)
2.	a	(149)
3.	b	(149)
4.	d	(152-153)
5.	c	(151-152)
6.	c	(152-153)
7.	c	(158)
8.	a	(159-160, 162)
9.	a	(160)
10.	a	(164)
11.	d	(164)
12.	d	(166)
13.	c	(168)
14.	b	(168)
15.	a	(170)

Matching: Definitions

1.	e	(148)
2.	p	(164)
3.	i	(150)
4.	h	(152)
5.	l	(159-160)
6.	a	(148-149)
7.	k	(157)
8.	o	(166)

9.	j	(165)
10.	c	(166)
11.	g	(154)
12.	n	(164)
13.	m	(152)
14.	f	(148)
15.	d	(149-150)
16.	b	(158)

Completion

1. target-seeking or self-stabilizing (149)
2. $2.5 billion a year (155)
3. subcortical (152)
4. right hemisphere; left hemisphere (153)
5. cephalocaudal (157)
6. hitching (160)
7. autoimmune; mathematical (164)
8. and; inside (168)
9. hypothesis (168)
10. biologically prewired (169-170)

Chapter 7

Infancy: Cognitive and Language Development

INTRODUCTION

Chapter 7 explores the processes by which cognition and language develop during early childhood. Cognitive and language abilities are the most distinctive features of human beings; without them, we would be without culture. The chapter concentrates on four main topics, including:

- Cognitive Development. According to Piaget's sensorimotor period and as formulated by various psychologists stimulated by Piaget's work, individual differences in mental performance throughout infancy are shown to be developmentally continuous across childhood and even into adulthood.

- Language and Thought. The functional importance of language is discussed, with the emphasis on the role of language in facilitating communication as well as thought. The relation of language to thought is examined in accordance with two opposing viewpoints: language as the mere container of thought and language as the determinant of thought.

- Theories of Language Development. Examined are the learning and interactionist theories, which state that language is acquired through learning processes, as well as the opposing innateness theory, which views human beings as biologically "prewired" for language usage. Chomsky's nativist theory, whereby the basic structure of language is biologically programmed, is presented in detail.

- Language Development. The importance of early nonverbal behaviors, prelanguage, language production, and language reception in the communication process is discussed. The sequence of language development in children is examined, beginning with such early vocalizations as cooing and babbling, progressing into holophrastic speech, two-word utterances, and telegraphic speech.

CHAPTER OUTLINE

I.   Cognitive Development

   A. Piaget: The Sensorimotor Period

   B. Neo- and Post-Piagetian Research

   C. Jerome S. Bruner on Modes of Cognitive Representation

   D. Continuity in Cognitive Development from Infancy

II. Language and Thought

   A. The Functional Importance of Language

   B. Language as the Container of Thought

   C. Language as a Determinant of Thought

III. Theories of Language Acquisition

   A. Learning and Interactionist Theories

   B. Innateness Theory

   C. A Resolution of Divergent Theories?

IV. Language Development

   A. Communication Processes

   B. The Sequence of Language Development

LEARNING OBJECTIVES

After completing Chapter 7, you should be able to:

1. List the three main characteristics of Piaget's *sensorimotor* period as they relate to cognitive development during the first two years of life. (pages 176-178)

2. Identify and explain Bruner's three modes of cognitive representation. (179)

3. Explain how the following two components can be used to predict cognitive competence in childhood: (180)

   a. decrement of attention

   b. recovery of attention

4. Discuss both the positive and the negative effects the home environment can have on the infant's cognitive development. (Box 181)

5. Define *language* and discuss some of the similarities between *language* and cognition. (181-182)

6. State the two vital contributions which *language* makes to the human condition, and explain the functional importance of each. (182)

7. Distinguish between the "*language* as container of *thought*" and the "*language* as a determinant of *thought*" positions. (183-185)

8. Discuss the learning and interactionist theories of *language* acquisition, and explain the role of *caretaker speech* and *motherese* in *language* development. (185-187)

9. Describe the innateness theory of *language* acquisition and discuss some of the major inadequacies which its proponents find in the position of the learning and interactionist theorists. (187-188)

10. Cite and critically evaluate research data regarding the creation of sign language by deaf infants. (Box 189)

11. Discuss Chomsky's theory of *language* development, specifically defining each of the following terms: (188-190)

    a. *language acquisition device*

    b. phonemes

    c. surface structure

    d. deep structure

    e. transformational grammar

12. Explain briefly the interplay of biological and environmental factors in the development of *language*. (190-191)

13. Assess the importance of the following terms in the *communication* process: (191-192)

    a. body language

    b. paralanguage

    c. language production

    d. language reception

14. Describe the following features of early *language* development: (192-196)

    a. crying

    b. cooing and babbling

    c. *holophrastic* speech

    d. overgeneralization

    e. two-word utterances

    f. telegraphic speech

caretaker speech   (page 186)

communication   (182)

conceptualization   (184)

holophrase   (194)

language   (182)

language acquisition device (LAD)   (189)

motherese   (186)

object permanence   (178)

sensorimotor   (177)

thought   (182)

SELF-TESTS - CHAPTER 7

Multiple-Choice

Circle the letter of the response which *best* completes or answers each of the following statements and questions.

Factual Questions:

1.   The study of cognition includes all of the following *except*

   a.   perception
   b.   problem solving
   c.   recall
   d.   reflexive reasoning

2.   The realization that objects continue to exist even when they are out of sight constitutes

   a.   symbolic representation
   b.   cognition
   c.   object permanence
   d.   object identity

3. According to Jerome Bruner, what type of cognitive representation appears first in children?

   a. ikonic
   b. enactive
   c. symbolic
   d. perceptual

4. Jerome Bruner indicates that children use mental images or pictures for representing the world. He refers to these as

   a. symbols
   b. enactive images
   c. ikonic images
   d. sign images

5. Which of the following statements regarding the prediction of intelligence from cognitive performance in infancy is the most accurate?

   a. Modern psychologists believe that there is no continuity between early and later capabilities.
   b. Decrement and recovery of attention seem most indicative of intelligence in youngsters.
   c. Youngsters who tire more easily when looking at one object are less efficient processors of information.
   d. Youngsters who prefer familiar stimuli over novel stimuli are more efficient processors of information.

6. In this chapter, language is defined as

   a. a structure system of sound patterns that have socially standardized meanings
   b. a set of rules for joining words to form phrases and sentences
   c. the process by which people transmit information
   d. a system of categories used to organize perceptual input

7. Proponents of the view that language is a container of thought claim that

   a. language precedes thought
   b. thought is not possible without language
   c. language shapes thought
   d. thought takes place independently of language

8. Children approach the task of "word learning" equipped with preexisting _____ that limit the number of possible meanings for a new noun.

   a. cognitive biases
   b. conceptualizations
   c. language biases
   d. symbols

9. A psychologist who accepted the view that language determined thought would tend to emphasize the importance of

   a. paralanguage
   b. ikonic representation
   c. caretaker speech
   d. conceptualization

10. Which of the following statements regarding color organization in infants is the most accurate?

    a. Color organization in infants precedes learned categories (i.e., "red," "blue") provided by language.
    b. Color organization is a product of the categories (i.e., "red," "blue") provided by language.
    c. Children less than one year of age lack the ability to partition the color spectrum into the four basic hues.
    d. Infants apply verbal labels to the categories (i.e., "red," "blue") before they have a mental representation of colors.

11. Interactionist theory is to _____ as innateness theory is to _____.

    a. nurture; nature
    b. learning; environment
    c. nature; nurture
    d. biological endowment; environment

12. In _____, parents systematically modify the language that they employ with adults when addressing young children.

    a. motherese
    b. caretaker speech
    c. conceptualization
    d. phonology

13. Nativists contend that the primary factor influencing language acquisition is

    a. the child's environment
    b. the interaction between the child and his or her caretaker
    c. learning processes
    d. the child's biological endowment

14. Which of the following is a nativist criticism of learning theory views of language acquisition?

    a. Children's speech is not a mechanical playback of adult speech.
    b. Caretakers spontaneously develop a unique language system for communicating with children.
    c. Children understand only about one-half the words spoken to them in motherese.
    d. The specialized anatomy of the vocal tract makes language possible.

15. Research on deaf infants has revealed all of the following *except* that

    a. when playing alone, deaf children employ signs of "talk" to themselves
    b. most of the deaf child's communication system is invented by the parents
    c. deaf infants can invent their own sign language of stereotyped gestures
    d. deaf babies string together signs and motions in much the same manner that hearing youngsters string together sounds

16. Noam Chomsky pointed out that the world's languages have basic similarities in their composition, and he termed these similarities

    a. transformational grammar
    b. deep structure
    c. surface structure
    d. caretaker speech

17. Most psychologists agree that there is a _____ for language acquisition, but they disagree over the importance of _____.

    a. biological basis; environmental factors
    b. cultural basis; genetic factors
    c. cultural basis; parental input
    d. biological basis; cognitive factors

18. The stress, pitch, and volume of words by which we communicate expressive meaning are referred to as

    a. language production
    b. kinesics
    c. paralanguage
    d. telegraphic speech

19. When a child uses the same single word and conveys different meanings, depending on the context he is in, the child is using

    a. holophrasic speech
    b. transformational grammar
    c. symbolic syntax
    d. deep structures

Conceptual Questions:

1. A 6-month-old child smiles and grows excited when she gazes at the teddy bear in her crib. However, when the infant's mother accidentally covers the bear with a blanket, the child becomes upset because she fails to realize that the bear still exists. According to Piaget, the child is demonstrating

    a. thought without cognition
    b. a learning disability
    c. lack of object permanence
    d. mental representation

2. An infant sees a ball under a chair; but when he crawls over to the side of the chair, the ball is not there. The infant is soon distracted by another toy. Psychologists who have revised Piaget's insights would say that this infant

    a. knew the ball still existed
    b. did not attribute permanence to objects
    c. had difficulty coordinating seeing and reaching
    d. did not possess cognitive representation

3. A little girl is playing with her doll, and the feet of the doll get caught in the neck of the little girl's dress. The little girl then bends her own fingers at a 90-degree angle and stares at them. This behavior indicates

    a. object permanence
    b. categorical concepts
    c. enactive representation
    d. symbolic representation

4. Bridget's baby quickly becomes drowsy and inattentive when she rocks her or lays her down on the floor to watch television. However, the baby quickly becomes reanimated when an interesting sight or sound occurs in the room. We might reasonably predict that Bridget's baby will

   a. become an impulsive and hyperactive child
   b. become a curious and effective problem solver
   c. not differ substantially from other children during preschool years
   d. not display categorical perception during her preschool years

5. Nathan, a 4-year-old, has been raised by overly ambitious parents who have pushed their son to gain academically oriented skills. We can reasonably infer that Nathan

   a. may grow to dislike learning as a result of the pressures to learn
   b. will be more apt to undertake highly skilled activities
   c. will develop a higher intellect than children with less demanding parents
   d. may benefit more from a structured preschool program than a culturally deprived youngster

6. Language is to _____ as cognition is to _____.

   a. perceptual input; semantics
   b. syntax; categorical relations
   c. phonology; perceptual input categories
   d. thinking; symbolism

7. A 2-year-old drinks from a cup without spilling even though she does not know the meaning of the word "tilt." However, a 6-year-old who knows the word "tilt" cannot accurately draw a picture of water in a tilting glass. This observation indicates that

   a. language and cognition are unrelated
   b. mental images of events may precede the use of linguistic representation
   c. categorical perception develops parallel with linguistic use
   d. language interferes with the acquisition of concepts

8. Nancy is a depressed mother who speaks to her infant using unexaggerated intonation contours of motherese. Moreover, she is slow to respond to her child's early attempts at vocalization. We can reasonably infer that Nancy's child

   a. will acquire language in much the same manner as the child of a healthy, happy mother
   b. will suffer a handicap in language acquisition
   c. will have an overdeveloped language acquisition device
   d. will remain unaffected by her mother's depression

9. In speaking to her 8-month-old baby, a mother uses shorter words, more pointing, more repetition, more exclamations (e.g., "gee," "gosh"), and more diminutives (e.g., "bunny" vs. rabbit) than when talking to her 5-year-old. This simplified speech

   a. is intended to develop the child's language acquisition device
   b. improves the child's object permanence ability
   c. attributes conceptual ability to the baby and by doing so facilitates language acquisition
   d. is pleasing to the parent but has no effect on the child because children this age have only productive speech

96

10. Your child uses word constructions such as "all gone sticky," "bye-bye hot," and "gooder." This type of language supports the

    a. learning theory position
    b. Piagetian position
    c. nativist position
    d. linguistic relativity position

11. The two sentences "Carl flew the kite" and "The kite was flown by Carl" have the same _____ but different _____.

    a. surface structures; deep structures
    b. semantics; phonology
    c. phonology; syntax
    d. deep structures; surface structures

12. Tony, who has normal hearing, lives with two deaf parents who communicate by sign language. Tony suffers severe asthma and is thus confined to his home where no English is spoken. By the time he is 4 years old, Tony

    a. will probably not understand nor speak English
    b. will not understand English or sign language
    c. will have learned English through daily television exposure
    d. will be fluent in both sign language and English

13. A child says the word "mama" when he wants to be fed. The child also uses the word "mama" to communicate that he wants to be let out of his crib. The child's use of the word "mama" could be described as a(n)

    a. overgeneralization
    b. body language
    c. transformational rule
    d. holophrase

14. A toddler uses sentences such as "go store" and "more cookie." This child is probably using

    a. general syntax rules
    b. telegraphic speech
    c. compound speech
    d. functional speech

15. A 2-year-old child says, "I falled down and hurted myself," and his mother corrects him by saying, "You mean 'I fell down and hurt myself.'" The child will probably reply:

    a. "I fell down and hurt myself."
    b. "You hurted yourself."
    c. "I falled down and hurted myself."
    d. "I fell down and hurted myself."

97

Matching: Definitions

Match the letter of the term in the box with its definition.

a.	cognition	i.	motherese
b.	sensorimotor	j.	language
c.	caretaker speech	k.	telegraphic speech
d.	language acquisition device	l.	conceptualization
e.	holophrase	m.	semantics
f.	object permanence	n.	communication
g.	thought	o.	syntax
h.	transformational grammar	p.	kinesics

1. the coordination of motor activities with sensory inputs _____

2. a simplified and redundant type of language employed by parents in communicating with their infant _____

3. the process of knowing _____

4. communication through body language _____

5. an inborn language-generating mechanism in all humans _____

6. the process whereby parents systematically modify the language that they employ with adults when addressing infants _____

7. single words that convey different meanings depending on the context in which they are used _____

8. grouping perceptions into classes or categories on the basis of certain similarities _____

9. the process by which people transmit information _____

10. rules for joining words to phrases and sentences _____

11. a system of meanings associated with words _____

12. a child's ability to view a thing as having a reality of its own that extends beyond the child's immediate perception of it _____

13. a structure system of sound patterns that have socially standardized meanings _____

14. Use of nouns and verbs in sentences, with the omission of prepositions, conjunctions, and articles _____

15. preverbal, intuitive rules which enable individuals to turn deep structure into surface structure and vice versa _____

16. the ability of an individual to represent and act mentally on absent objects and events _____

Completion

Supply the term of terms needed to complete each of the following statements.

1. A language _____ device is an inborn language-generating mechanism.

2. _____ was a cognitive theorist who assumed that infants cannot symbolically fashion mental representations of the world.

3. _____ mental images are closely related to perception.

4. The process of joining together units of sound to form words is called _____.

5. Babies' early squealing-gurgling noises are known as _____.

6. _____ speech is sometimes referred to as baby talk.

7. The grouping of perceptions into categories based on certain similarities is known as _____.

8. Children approach the task of "word learning" equipped with preexisting _____ that lead them to prefer some possible meanings over others.

9. _____, such as Chomsky, contend that human beings are "prewired" by their brain circuitry for language use.

10. The basic similarities in the composition of the world's languages are called _____.

Multiple-Choice

Factual

1.	d	(176)
2.	c	(177)
3.	b	(179)
4.	c	(179)
5.	b	(179-180)
6.	a	(182)
7.	d	(183)
8.	a	(183)
9.	d	(184)
10.	a	(184)
11.	a	(185, 187)
12.	b	(187)
13.	d	(187-188)
14.	a	(188)
15.	b	(189)
16.	b	(190)
17.	a	(190)
18.	c	(192)
19.	a	(194)

Conceptual

1.	c	(178)
2.	a	(179)
3.	c	(179)
4.	b	(180)
5.	a	(181)
6.	c	(182)
7.	b	(183)
8.	b	(187)
9.	a	(186-187)
10.	c	(187-188)
11.	d	(190)
12.	a	(193-194)
13.	d	(194)
14.	b	(196)
15.	c	(195-196)

Matching: Definitions

1.	b	(177)
2.	i	(186)
3.	a	(176)
4.	p	(191)
5.	d	(189)
6.	c	(186)
7.	e	(194)
8.	l	(184)

9.	n	(182)
10.	o	(182)
11.	m	(182)
12.	f	(178)
13.	j	(182)
14.	k	(196)
15.	h	(190)
16.	g	(183)

Completion

1.	acquisition (189)	
2.	Piaget (176)	
3.	ikonic (179)	
4.	phonology (182)	
5.	cooing (193)	

6.	caretaker (186)
7.	conceptualization (184)
8.	cognitive biases (183)
9.	nativists (187)
10.	deep structure (190)

Chapter 8

Infancy: Psychosocial Development

INTRODUCTION

Chapter 8 illustrates the importance of children's early years in shaping their psychological and social well-being. The chapter covers two main categories, including:

- The development of Emotion and Social Bonds. The role of emotion is discussed, and research pertaining to the emotional life of children is presented. Various theoretical perspectives attempt to explain the changes in the development of emotion. The objects, functions, and formation of the attachment process are illustrated, as well as the effect of early attachment patterns on later relationships.

- Early Parenting: Myths and Facts. An abundance of research reveals the effects of institutionalization on children. Early parenting practices, the role of the father, and multiple mothering are all discussed regarding the child's psychosocial development. The effects of day care on the child's emotional, social, and intellectual development are covered, with much research devoted to the problems in our current child-care institutions.

I. The Development of Emotion and Social Bonds

    A. The Role of Emotion

    B. Emotional Development in Infants

    C. Theoretical Perspectives on Emotional Development

    D. Attachment

    E. Stranger Wariness and Stranger Anxiety

II. Early Parenting: Myths and Facts

    A. Institutionalized Children

    B. How Important are Early Parenting Practices?

    C. The Role of the Father

    D. Multiple Mothering

    E. Child Day-Care Centers

LEARNING OBJECTIVES

After completing Chapter 8, you should be able to:

1. Define the terms *emotion* and *social referencing* and discuss some of the functions which are served by emotional reactions. (page 202)

2. Cite findings from the chapter on each of the following topics pertaining to child abuse:

    a. Identify the different types of child abuse and neglect. (Box, 204-206)

    b. Describe the characteristics of the victims of abuse and neglect. (Box, 204-206)

    c. Cite the characteristics of abusive and neglectful parents. (Box, 204-206)

3. Discuss the research data on emotional development in infants regarding each of the following: (203, 206-207)

   a. the role of the central nervous system in facial expression of emotion (203)

   b. the differential emotions theory (203)

   c. the continuity of emotional expression in children (206)

   d. the six stages in children's emotional development according to Stanley and Nancy Greenspan (207)

4. Compare and contrast the cognitive and the nativist approaches to emotional development. (208-209)

5. Answer the following questions regarding infant *attachment*: (209-213)

   a. How is specific *attachment* indicated? (209)

   b. Who are the objects of *attachment*? (210)

   c. What functions are served by *attachment*? (211)

   d. What is the predictive value of early *attachment* patterns? (213)

6. Describe the major differences between the learning theory and ethological views of the *attachment* process. (211-212)

7. Briefly summarize the research data on the development of *stranger anxiety* in children. (213-214)

8. Describe the concept of *maternal deprivation*, and discuss the effects of institutionalization on children. (214)

9. Discuss the current crisis in foster care, and summarize the pros and cons associated with orphanages. (217)

10. Describe *developmental psychology*, and examine its significance. (219)

11. Briefly discuss Freud's views on development during infancy, and evaluate Stern's conflicting research on early parenting practices. (212, 216, 218)

12. Examine the role of the father in the development of young children. (220-221)

13. Describe *multiple mothering*, and examine its effects on young children. (224)

14. Critically appraise the effects of day care on young children. (224-227)

attachment (page 209)

developmental psychopathology (Box, 219)

emotion (202)

maternal deprivation (214)

multiple mothering (224)

person permanence (210)

social referencing (202)

stranger anxiety (213)

strange situation (212)

SELF-TESTS - CHAPTER 8

Multiple-Choice

Circle the letter of the response which *best* completes or answers each of the following statements and questions.

Factual Questions:

1. Child neglect is best defined as

    a. intentional physical attack on a child by the parent
    b. injury that is purposely inflicted on the child by the parent
    c. absence of adequate social, emotional, and physical care
    d. severe injuries associated with physical abuse of children

2. Studies of parents who abuse their children reveal that the parents

    a. expect very little from their children
    b. are exclusively from lower socioeconomic levels
    c. were themselves abused as children
    d. usually suffer from severe psychotic tendencies, untreatable through counseling

3. Which of the following statements is the *most* accurate according to Izard's differential emotions theory in infants?

   a. All the basic emotions are in place at birth.
   b. At birth, the inner feelings of babies are limited to shyness, contempt, and guilt.
   c. Emotions are not preprogrammed on a biological clock.
   d. Each emotion has its own distinctive facial pattern.

4. According to the nativist view,

   a. emotions are products of developing thought processes
   b. emotional expressions cannot operate independently of thought
   c. noncognitive factors play a primary role and cognition only a secondary role in the development of emotion
   d. the amygdala triggers an emotional reaction after the thinking brain has fully processed nerve signals

5. Which of the following could we say was an indicator of specific attachment in infants?

   a. separation distress
   b. bonding
   c. maternal deprivation
   d. developmental psychopathology

6. Research by Schaffer suggests that the onset of separation distress is directly related to a child's level of

   a. neglect
   b. object permanence
   c. attachment
   d. stranger wariness

7. Which of the following is *not* a function of attachment?

   a. The child learns to cling to the primary caregiver.
   b. The child learns about the world.
   c. It provides the child with emotional security.
   d. It allows the child to enter into social relationships with other human beings.

8. According to learning theories of attachment, the mother is initially a

   a. biologically preprogrammed magnet for the child
   b. punishing source for the child
   c. hindrance to the child's acquisition of gender identity
   d. neutral stimulus for the child

9. Research by Ainsworth indicates that

   a. there is no correlation between early maternal caregiving and patterns of attachment behavior
   b. the avoidant mothers of resistant infants are often inconsistent, insensitive, and rejecting
   c. attachment patterns do not vary from culture to culture
   d. early attachment behaviors are not indicative of later social and cognitive development

10. In which of the following situations are babies most wary of strangers?

    a. when their mothers are not present
    b. when they are confronted by a child stranger
    c. when a stranger approaches them
    d. when a stranger picks them up in the child's own home

11. The view that the absence of normal mothering can result in psychological damage and physical deterioration in children finds expression in

    a. social deprivation
    b. maternal deprivation
    c. institutionalization
    d. sensory deprivation

12. Research by Wayne Dennis on the effects of institutional life on infants suggests that

    a. institutionalized children suffer developmentally because of solely the absence of the mother-child tie
    b. the absence of adequate cognitive experiences is the primary reason institutionalized children develop inadequately
    c. after 2 years of age, the effects of deprivation caused by inadequate cognitive experience are always reversible
    d. children reared in institutions are more socially and intellectually developed than children reared in foster homes

13. Which of the following factors has *not* contributed significantly to the current crisis in foster care?

    a. the crack epidemic
    b. inadequate payments to foster parents
    c. rising numbers of women entering the job market
    d. an overabundance of youngsters being adopted into permanent families

14. Developmental psychopathologists, in studying the emotional problems of children, find that

    a. children with emotional problems tend to have fewer psychiatric difficulties in adulthood
    b. factors such as age, sex, and ethnic background do not contribute to rates of childhood dysfunction
    c. children who function poorly in an academic environment may perform better in the home or community
    d. approximately one in twenty youngsters is affected by developmental, learning, and behavioral disorders

15. Current research on the role of the father indicates that

    a. men lack the potential to be as good caretakers of children as women
    b. fathers are more likely than mothers to hold their babies and look at them
    c. fathers smile more at their babies than mothers
    d. babies drink less milk when fathers do the bottle feeding than when mothers do

16. According to Michael E. Lamb, fathers are more likely to hold babies to _____, whereas mothers are more likely to hold their infant to _____.

    a. play with them; perform caretaking functions
    b. punish them; play with them
    c. play with them; punish them
    d. perform caretaking functions; play with them

Conceptual Questions:

1. Roxanne's 1-year-old son looks at her for guidance before he allows a photographer to take his picture. When his mother smiles at him, the child remains at ease until the camera flicks. Roxanne's son has demonstrated

   a. specific attachment
   b. person permanence
   c. social referencing
   d. stranger anxiety

2. A mother tells a therapist, "My 3-month-old does not look at me. I think she doesn't love me." Such a misreading indicates that the

   a. child has probably developed maternal deprivation syndrome
   b. mother may abuse the child when the child doesn't satisfy her needs
   c. mother has psychotic tendencies and should be institutionalized
   d. child is demonstrating specific attachment behavior

3. A child in your first-grade class arrives at school fifteen minutes early and wants to stay after school to help you clean up the room. This behavior may be

   a. the way the child deals with separation anxiety
   b. an indication of secure attachment to the mothering figure
   c. an indicator of recent maternal deprivation
   d. a potential warning sign of child abuse and neglect

4. Your nephew is 8 months old. You are a psychology major interested in assessing his emotional development. You might expect that your nephew will

   a. lift his arms when his mother bends over to pick him up and gurgle when his mother says, "Hi"
   b. primarily display either surprise or sadness, particularly if he has never seen you
   c. respond in a stereotypical way toward you even when you change your facial expressions
   d. approach you if you raise your eyebrows and wrinkle your nose .

5. An infant is more likely to approach, follow, and cling to its mother than to its father. These behaviors indicate

   a. unfocused expression and excitement
   b. a specific attachment
   c. an insecure attachment
   d. arousal by all parts of the environment

6. You are concerned because your 8-month-old does not seem to feel distressed when you leave the house. Your sister, who is a full-time homemaker, says that her children showed separation anxiety at 6 months of age. On the basis of research evidence, you might conclude that

   a. because you work, separation is a more common event for your infant, and thus he reacts less to it
   b. your child has probably developed person permanence, whereas your nephews have not
   c. your nephews are showing indiscriminate attachment, whereas your child is showing specific attachment
   d. your child has not been imprinted by a social network of other adults

108

7. Which of the following statements is consistent with the learning-theory view of attachment?

   a. Behaviors that promote proximity are important in the natural selection process.
   b. Attachment develops through the process of socialization, whereby children acquire a need for the presence of the parents.
   c. Parents are genetically predisposed to provide caretaking behavior.
   d. The human face is an innate releasing stimulus for smiling.

8. Stephanie and her mother are visiting a friend for the first time. Stephanie plays happily until her mother leaves to get some coffee. She starts to cry but quickly stops when her mother returns. What type of attachment does she have with her mother?

   a. secure
   b. avoidant
   c. resistant
   d. ambivalent

9. Pedro is a 1-year-old who is securely attached to his mother. Which of the following statements is the *least* accurate?

   a. He has received consistent, sensitive, responsive mothering.
   b. Pedro is likely to develop into an unselfish toddler and will acquire normal social skills.
   c. He is likely to develop good cognitive skills.
   d. As he ages, Pedro will lack the ability to care for his younger siblings because he demands too much attention from his mother.

10. Dana is out walking with her father when a woman she does not know approaches them rapidly. Dana starts to cry. Her crying probably signifies

    a. wariness of strangers
    b. attachment to her mother
    c. separation distress
    d. attachment to her father

11. Chelsea is 11 months old when her mother drops her off at the local day-care center for the first time. She soon exhibits friendly responses toward the new caretaker. This behavior

    a. is abnormal, because most infants are afraid of strangers in all new situations
    b. is normal, because most infants find strangers to be the most comforting object in an alien environment when their mothers are absent
    c. indicates that Chelsea has not yet embarked upon the developmental stage referred to as "fear of the stranger"
    d. is normal, because most children Chelsea's age prefer to be in the presence of strangers than with their mother

12. Alex was raised in an institution until the Romanos adopted him at the age of 4. According to research evidence by Wayne Dennis, we can assume that

    a. Alex will experience no mental deficiency whatsoever
    b. although Alex will attain a higher IQ score after adoption, he will retain a certain amount of mental deficiency resulting from his institutionalization
    c. the effects of the institutionalization are reversible, and Alex will attain a normal IQ score after he is reared in a normal family environment
    d. Alex's intelligence level will remain unaffected by the processes of institutionalization and adoption

13. Your 5-year-old son displays distress when you leave him at the day-care center. Developmental psychopathologists would probably say that

   a. your son will be more likely than other children to have psychiatric difficulties as an adult
   b. your son was displaying the early precursors of a bipolar mood disorder
   c. this behavior was an ethologically significant and normal part of your son's attachment behavioral system
   d. the duration, frequency, and intensity of his behavior would need to be assessed before its significance could be determined

14. When Jody's baby boy squeals in delight, she gives him a gentle shake. According to Daniel Stern, such attunements that evolve between children and their caretakers

   a. give infants a reassuring sense of being emotionally connected to other people
   b. do not have long-term consequences on the child's future development
   c. cause children to develop deficiencies in their social development
   d. support the Freudian view that early childhood experiences determine later personality characteristics

15. A parent kisses, touches, and plays rough-and-tumble games with an infant son. Which of the following conclusions is the most reasonable?

   a. This boy will be less affected by the absence of his parent than a daughter would be.
   b. This parent is most likely to be a mother.
   c. This parent is most likely to be a father.
   d. This parent will not respond as quickly to the infant's distress signals.

16. Jason is a 5-year-old child afflicted by childhood autism. According to research by the journalist Richard Louv, which of the following statements is the most accurate?

   a. Jason's disorder is most likely attributable to the cold, rejecting behavior of his mother.
   b. Biological and genetic determinants play a large role in Jason's disorder.
   c. With the proper amount of motherly love and attention, Jason will overcome his disorder.
   d. Jason's autism is a result of neglectful and unmotivated fathering.

17. A 10-month-old child responds aggressively to adults and other children. On the basis of research cited in your text, a child development expert could recommend that enrolling this child in a day-care center might

   a. aggravate his aggressive tendencies
   b. increase his cooperation and decrease his aggressiveness
   c. increase his social skills if there are many other children but few adult supervisors
   d. improve his cognitive development but not affect his social skills

Matching: Definitions

Match the letter of the term in the box with its definition.

a.	stranger anxiety	i.	attachment
b.	multiple mothering	j.	strange situation
c.	neglect	k.	social network
d.	emotion	l.	abuse
e.	developmental psychopathology	m.	indiscriminate attachment
f.	person permanence	n.	Sigmund Freud
g.	maternal deprivation	o.	specific attachment
h.	social referencing	p.	Daniel Stern

1.   an affectional bond enduring across time and space _____

2.   an inexperienced person relies on a more experienced person's perception of an event _____

3.   the second stage in the development of infant social responsiveness _____

4.   believes that relationships throughout life continually shape and reshape our working schemes or models of social relationships _____

5.   feelings associated with psychological changes, subjective experiences, and expressive behavior _____

6.   psychological and physical retardation resulting from absence of normal mothering _____

7.   a baby's preference for a particular person _____

8.   idea that a person exists even if not immediately visible _____

9.   stressed that adult neurosis has its roots in childhood conflicts associated with the meeting of instinctual needs _____

10. wariness of strangers that peaks around 10 months of age _____

11. dispersion of responsibility for a child's care among several people _____

12. an encompassing web of ties to significant others _____

13. procedure used to determine whether infants are attracted to their parents in a secure, avoidant, or resistant manner _____

14. psychologists interested in this focus on the processes explaining how early patterns of adaptation evolve into later adaptation patterns _____

15. absence of adequate social, emotional, and physical care _____

16. nonaccidental physical attack on or injury to children by individuals caring for them _____

Completion

Supply the term or terms needed to complete each of the following statements.

1. Erikson maintains that the essential task of infancy is the development of a basic _____ in others.

2. Families that are socially _____ and outside neighborhood support networks are more likely to abuse children than families with rich social ties.

3. First, emotions seem to have evolved as _____ processes that enhance survival.

4. Emotional expressions and behaviors in infancy are related to _____ characteristics at 5 years of age.

5. When infants seek arousal equally from human and nonhuman aspects of their environment, we describe them as displaying _____ _____.

6. _____ psychologists emphasize the adaptive value of attachment in promoting proximity between infants and caretakers.

7. Babies who, upon their mothers' return to the strange situation, mixed active control seeking with squirming were described as being _____ by Ainsworth.

8. Recently, researchers have discovered that physical contact and sensory stimulation serve to improve the sensorimotor functioning of children in _____ settings.

9. Fathers' rights advocates rightfully assert that physiological responses human beings display to infant coos and other signals seem to be _____.

10. For economically disadvantaged children, the _____ experience can lessen the decline in IQ scores that typically takes place among children from "high-risk" environments.

Multiple-Choice

Factual

1.	c	(204)
2.	c	(205)
3.	d	(203)
4.	c	(208)
5.	a	(209)
6.	b	(210)
7.	a	(211)
8.	d	(212)
9.	b	(212)
10.	a	(213)
11.	b	(214)
12.	b	(215)
13.	d	(217)
14.	c	(219)
15.	b	(221)
16.	a	(221)

Conceptual

1.	c	(202-203)
2.	b	(204)
3.	d	(205)
4.	a	(207)
5.	b	(209)
6.	a	(209-210)
7.	b	(212)
8.	a	(212)
9.	d	(213)
10.	a	(213)
11.	b	(214)
12.	b	(216)
13.	d	(219)
14.	a	(218, 220)
15.	c	(221)
16.	b	(222)
17.	a	(226)

Matching: Definitions

1.	i	(209)
2.	h	(202)
3.	m	(209)
4.	p	(220)
5.	d	(202)
6.	g	(214)
7.	o	(209)
8.	f	(210)

9.	n	(216)
10.	a	(213)
11.	b	(224)
12.	k	(210)
13.	j	(212)
14.	e	(219)
15.	c	(204)
16.	l	(204)

Completion

1.	trust (200)
2.	isolated (204)
3.	adaptive (202)
4.	personality (206)
5.	indiscriminate attachment (209)

6.	ethological (211-212)
7.	resistant (212)
8.	institutionalized (216)
9.	species specific (220)
10.	day care (225-226)

Chapter 9

Early Childhood: Expanding Competencies

INTRODUCTION

Whereas Chapters 6, 7, and 8 dealt with the period of infancy, Chapter 9 is the first of three chapters focusing on early childhood--the time between ages 2 and 6. During this period, children acquire autonomy, evolve new ways of relating to other people, and gain a sense of their effectiveness in the world. Several vital topics are examined, including:

- Cognitive Development. Various forms of intelligence are discussed, as is the nature-nurture controversy surrounding the level of an individual's intelligence. The early development of intelligence is traced in accordance with Piaget's observations. More recent research concerning children's conceptual foundations for learning is also presented.

- Information Processing and Memory. Early memory, information processing, metacognition, and metamemory are discussed in detail. The memory strategies of categorizing and rehearsing are illustrated in this section as well.

- Cognitive Foundations for Social Interaction. The ways in which children acquire conceptions of role and self are discussed. Roles provide social guidelines which define an individual's obligations as well as expectations. The self provides an individual with the capacity to observe, respond to, and direct his or her behavior.

- Gender Identification. The hormonal and social influences on gender behaviors are discussed. Three theories regarding the acquisition of gender identities are compared and contrasted. Also examined are the stereotypes parents have concerning the behaviors they expect to be associated with male infants and female infants.

CHAPTER OUTLINE

I. Cognitive Development

    A. Intelligence

    B. Intelligence and the Nature-Nurture Controversy

    C. The Early Development of Intelligence

    D. Conceptual Foundations for Learning

II. Information Processing and Memory

    A. Early Memory

    B. Information Processing

    C. Metacognition and Metamemory

    D. Categorizing as a Memory Strategy

    E. Rehearsing as a Memory Strategy

III. Cognitive Foundations for Social Interaction

    A. Roles

    B. The Self

IV. Gender Identification

    A. Hormonal Influences on Gender Behaviors

    B. Social Influences on Gender Behaviors

    C. Theories Regarding the Acquisition of Gender Identities

    D. Mothers, Fathers, and Gender Typing

LEARNING OBJECTIVES

After completing Chapter 9, you should be able to:

1. Define *intelligence*, and in the context of this definition, discuss the following: (pages 232-236)

    a. *intelligence* as a general factor or composite of abilities

b.   the different kinds of *intelligence*

c.   *intelligence* as information processing

2.   Contrast *intelligence* and creativity, and describe the conditions that facilitate creativity. (234-235)

3.   Critically evaluate the IQ and race controversy. (236-237)

4.   Formulate a nature and a nurture position with regard to the determination of *intelligence*, and describe the research methodologies and types of research data you would use to defend your positions. (237-239)

5.   Characterize the thinking of the preoperational child by describing: (240-243)

a.   symbol use

b.   *conservation* performance

c.   *centering*

d.   state vs. transformational reasoning

e.   nonreversibility

f.   egocentrism versus sociocentrism

6. Discuss the conceptual foundations for learning in young children with regard to both *causality* and number concepts. (243-245)

7. Describe the following *memory*-related phenomena: (245-248)

   a. early *memory*

   b. types of *memory*

   c. *metacognition*

   d. *metamemory*

   e. *memory* strategies

   f. *short-term memory*

8. Discuss the evolution of the child's knowledge regarding social *roles*, decision making, and *self*-awareness. (248-252)

9. Explain the distinction between *gender roles* and *gender identities*, and assess the *role* that hormonal influences, brain structure, and social influences play in gender behavior. (253-256)

10. Summarize and critically evaluate the research of Maccoby and Jacklin on sex differences. (254-255)

11. Describe and evaluate the following theories of *gender identity*: (257-258)

    a. psychoanalytic theory

    b. cognitive learning theory

    c. cognitive-developmental theory

12. Appraise the research connected with the effects of fathers and mothers on gender typing. (259-261)

KEY TERMS

---

androgyny (page 261)

causality (244)

centering (241)

conservation (240)

correlation coefficient (238)

egocentrism (242)

gender identity (252)

gender role (252)

hermaphrodite (256)

intelligence (233)

long-term memory (247)

memory (245)

metacognition (247)

metamemory (247)

preoperational period (240)

reversible (242)

role (248)

self (250)

sensory information storage (246)

short-term memory (246)

two-factor theory of intelligence (233)

SELF TESTS - CHAPTER 9

Multiple-Choice

Circle the letter of the response which *best* completes or answers each of the following statements and questions.

Factual Questions:

1. The best way to describe intelligence is to call it

   a. the possession of a fund of knowledge
   b. a capacity for acquiring knowledge and functioning rationally and effectively
   c. a type of metamemory
   d. the retention of what has been experienced

2. Psychologists who view intelligence as a process are not so much interested in
   _____ we know, but in _____ we know.

   a. what; how
   b. why; how
   c. how; what
   d. what; when

3. Which of the following statements is the most accurate regarding the association between intelligence and creativity?

   a. High intelligence guarantees creative activity.
   b. Low intelligence seems to enhance creative activity.
   c. Formal education is essential for creativity.
   d. The optimal IQ for creativity is about nineteen points above the average of people in a given field.

4. Which of the following recommendations could be used to encourage creative thinking in children?

   a. Provide the child with problem-solving metacomponents.
   b. Afford the child opportunities to engage in associative learning.
   c. Confront the child with contradictions and uncertainties.
   d. Have the child reproduce artistic models you have created.

5. Arthur Jensen has used _____ studies to support his hereditarian position on the nature of racial differences in intelligence.

   a. family resemblance
   b. correlational kinship
   c. gene-environment covariance
   d. genetic experiential

6. If intelligence is primarily determined by genetic factors, then

   a. fraternal twins will be more similar to one another than regular siblings
   b. fraternal twins raised in differing environments would be more similar to one another than regular siblings
   c. identical twins who were highly intelligent would also be highly creative
   d. little could be done to improve people's abilities through education, and social arrangement would be unchangeable

7. Sociologist Cristopher Jenecks has introduced the third element of gene-environment covariance to the nature-nurture controversy. This added component of variance

   a. is associated primarily with genetic factors
   b. is associated primarily with environmental factors
   c. is a result of the association of the separate effects of nature and nurture
   d. is a measure of family resemblance

8. Children first develop the capacity to represent the external world internally through symbols during the _____ period.

   a. preoperational
   b. sensorimotor
   c. concrete operational
   d. formal operational

9. The concept that the quantity or amount of something stays the same regardless of changes in its shape or position is called

   a. conservation
   b. transformation
   c. roles
   d. centering

10. Recent research has questioned Piaget's theory that preoperational children are completely
_____.

    a. altruistic
    b. sociocentric
    c. egocentric
    d. nonconservative

11. Contemporary developmental psychologists, in measuring children's counting capabilities, would argue with Piaget's viewpoint that

    a. there is no connection between the acquired ability to count and the actual operations of which the child is capable
    b. preschoolers have an implicit understanding for all number concepts
    c. counting is an "easy" cognitive accomplishment for all young children
    d. young children seem to possess some basic knowledge of "quantity" before they could have acquired such knowledge from their learning experiences

12. We experience a feeling of familiarity when we again perceive something that we have previously encountered. Psychologists call this

    a. recall
    b. rehearsal
    c. recognition
    d. facilitation of relearning

13. Information from the senses is preserved just long enough to permit the stimuli to be scanned for processing. This provides a relatively complete, literal copy of the physical stimulus and best describes

    a. short-term memory
    b. sensory information storage
    c. long-term memory
    d. rehearsal

14. The system of concepts we use in defining ourselves is best termed

    a. the self
    b. expectations
    c. social categories
    d. role

15. The self-conceptions that people have of themselves as being male or female are called

    a. gender roles
    b. genetic genders
    c. biological sexes
    d. gender identities

16. Maccoby and Jacklin studied sex differences between boys and girls. They found that around age 11, girls begin to

    a. move ahead of boys in mathematical ability
    b. exhibit greater verbal ability than boys
    c. fall behind boys in tasks requiring analysis
    d. exhibit less self-esteem than boys

17. Individuals having the reproductive organs of both sexes are called

   a. homosexuals
   b. transvestites
   c. transsexuals
   d. hermaphrodites

18. "I am a boy; therefore, I want to do boy things; therefore, the opportunity to do boy things is rewarded." This best describes the position of which of the following theories?

   a. psychoanalytic
   b. social modeling
   c. cognitive-developmental
   d. none of the above

19. The term commonly used to refer to the capacity of men and women to be both masculine and feminine in their attitudes and behavior is

   a. androgyny
   b. transvestism
   c. reversibility
   d. hermaphroditism

Conceptual Questions:

1. You give your third-grade students a battery of aptitude tests. Their performance on each test is highly correlated with their performance on all their other tests. You might therefore conclude that your test battery

   a. successfully tapped into the "g" factor
   b. successfully tapped into the "s" factor
   c. was not valid because it measures only composite factors
   d. measured primarily mental abilities

2. A parent who wants to raise a creative child should

   a. make sure the child has a high IQ
   b. read a lot of books to the child
   c. allow the child to initiate his or her own learning effort
   d. make sure the child receives a formal education

3. According to the hereditarian position, we would expect the correlation between the IQ scores of two unrelated children living in the same home to be

   a. higher than the correlation of fraternal twins raised in separate homes
   b. about the same as that of unrelated children raised in separate homes
   c. very high, because the same environment causes people to be similar
   d. unpredictable because we know nothing of their biological backgrounds

4. The average IQ of black children is generally ten to fifteen points below that of whites. However, Latonya, a black girl raised by white adoptive parents, achieved an IQ score much closer to the white norm. This finding supports

   a. Jensen's associative model of intelligence
   b. an environmental model of intelligence
   c. a family resemblance model of intelligence
   d. a hereditarian model of intelligence

5. Stephen claims that his mother gave his brother a bigger piece of cake than he got. However, Stephen is pacified when his mother cuts his piece in two. He defiantly says to his brother, "Now I've got more." Stephen's statement shows

    a. good perspective-taking ability
    b. sociocentric behavior
    c. associative learning but no conceptual learning
    d. lack of conservation

6. A young child sees two rows of pennies and says that each row contains five pennies. When the bottom row is spread out, she says that it has more pennies, while she simultaneously gestures with her hands spread apart. The hand gesture could be considered an example of

    a. centering
    b. encoding
    c. reversibility
    d. metamemory

7. Stephanie, a girl of 18 months, sees her mother lying down to rest. Stephanie goes to her crib, collects her favorite blanket, and covers her mother with it. Stephanie is demonstrating

    a. sociocentrism
    b. egocentrism
    c. centering
    d. typical Piagetian behavior

8. Many current developmental psychologists disagree with Piaget's views on the limited cognitive abilities of preschool children. According to these contemporaries, which of the following statements would a typical 4-year-old make?

    a. "The street makes the car go."
    b. "My teddy bear can talk to me."
    c. "The doll cannot walk by herself."
    d. "The blanket is tired and taking a nap."

9. A second-grade child realizes that she will have to spend more time studying words she has misspelled than words she can spell correctly. This child has a sophisticated

    a. sensory information storage
    b. recognition memory
    c. recall
    d. metamemory

10. A child is given a randomly arranged list of twelve words (e.g., names of animals, fruits, and colors). As the number of learning trials increases, he becomes more likely to recall all the animal names in a group. The child is using

    a. rehearsing
    b. metamemory
    c. a clustering form of categorization
    d. a serial ordering form of categorization

11. Eleanor and Victor have taught their two children to make their own decisions and solve their own problems. According to research by Elias and Clabby, we can predict that Eleanor and Victor's children

    a. will think independently and, thus, be more resistant to peer pressure
    b. will develop discipline problems
    c. will develop lower self-esteem than youngsters raised in authoritarian families
    d. none of the above

12. A 1-year-old baby points at a picture of her mother and says, "Mommy." When she sees a picture of herself, she says, "Baby." This behavior indicates

    a. lack of object permanence
    b. retarded use of self-referents
    c. fascination with the human face
    d. normal development of self-concept

13. Karen and Ricardo are middle school students who compete along the lines of "anything you can do, I can do better." Ricardo will probably be right if he compares his _____ performance with Karen's.

    a. history
    b. geometry
    c. English
    d. creativity

14. A mother is administered male hormones during pregnancy to prevent the miscarriage of her female fetus. On the basis of research by John Money, we can conclude that the mother's daughter will

    a. become a hermaphrodite
    b. prefer to play with dolls and show interest in cosmetics, jewelry, and perfume
    c. become a lesbian
    d. be more aggressive than the average female

15. Oedipal conflict is to psychoanalytic theory as _____ is to cognitive learning theory.

    a. gender role
    b. gender identity
    c. selective reinforcement
    d. self-socialization

16. Bill is a real macho-type, strong, and silent man. He and his wife have just brought their newborn son home from the hospital. Which of the following conclusions would be the most legitimate with regard to their son?

    a. His gender identity will be the same as Bill's.
    b. He will probably become an androgynous adult.
    c. He will be subjected to less punishment from his father than from his mother.
    d. His masculinity will not be affected by Bill's frequent absence from the home.

Matching: Definitions

Match the letter of the term in the box with its definition.

a.	androgyny	l.	preoperational period
b.	causality	m.	reversible
c.	centering	n.	role
d.	conservation	o.	self
e.	correlation coefficient	p.	sensory information storage
f.	egocentrism	q.	short-term memory
g.	gender identity	r.	two-factor theory of intelligence
h.	gender role	s.	intelligence
i.	hermaphrodite	t.	memory
j.	metacognition	u.	long-term memory
k.	metamemory		

1. a numerical expression indicating the degree of relationship between variables _____

2. retention of information for no more than thirty seconds _____

3. views intelligence as consisting of a "g" factor and an "s" factor _____

4. term used to describe operations capable of being turned back to an earlier state _____

5. conception of oneself as being male or female _____

6. cognitive stage (ages 2 to 7) when the ability to represent the world symbolically is acquired _____

7. system of concepts used to define ourselves, consisting of awareness of oneself as a separate entity _____

8. a global capacity to understand the world, think rationally, and cope resourcefully with the challenges of life _____

9. an individual with the reproductive organs of both sexes _____

10. attribution of cause-and-effect relationships to two paired events recurring in succession _____

11. retention of what has been experienced or learned _____

12. capacity of a person to be both masculine and feminine in attitudes and behavior _____

13. definition that specifies who does what and when and where they do it; also a set of behavioral expectations _____

14. cultural expectations defining appropriate behaviors for each sex _____

15. concentration on one feature of a situation while ignoring other aspects, characteristic of the preoperational stage _____

16. preserves sensory information long enough for it to be scanned and processed _____

17. lack of awareness of viewpoints other than one's own _____

18. awareness and understanding of one's own mental processes _____

19. the realization that quantity remains constant despite changes in shape or position _____

20. enables one to retain material over an extended period of time; depends on rehearsal or a very intense single experience _____

21. awareness and understanding of one's own memory processes _____

Completion

Supply the term or terms needed to complete each of the following statements.

1. Marco tells his mommy that daddy is no longer daddy after his father gets his long hair cut short. Marco is making a type of _____ error.

2. The optimal IQ for _____ is about nineteen points above the average of people in a given field.

3. A number of psychologists contend that _____ tests are biased in favor of white middle-class Americans.

4. _____ believe that intellect is increased or decreased according to the degree of enrichment or impoverishment provided by a person's social and cultural environment.

5. According to Piaget, the most important characteristic of preoperational thought is the child's failure to recognize that operations are _____.

6. Elias and Clabby believe that teaching children to make decisions and solve problems develops their capacity to think _____.

7. Roles impinge on us as sets of social norms that define our _____ and _____.

8. During the preschool years, young children conceive of the self strictly in _____ terms.

9. During the prenatal period, _____ suppresses the "female" pattern, so that "male" neural tissues become organized in programming later male sexual responses.

10. On the basis of his research with hermaphrodites, John Money concludes that the most powerful factors in the shaping of _____ _____ are environmental.

Multiple-Choice

Factual

1.	b	(233)
2.	a	(234)
3.	d	(234)
4.	c	(235)
5.	a	(237-238)
6.	d	(237)
7.	c	(240)
8.	a	(240)
9.	a	(240)
10.	c	(242)
11.	a	(245)
12.	c	(246)
13.	b	(246)
14.	a	(250)
15.	d	(252)
16.	b	(254)
17.	d	(256)
18.	c	(258)
19.	a	(261)

Conceptual

1.	a	(233)
2.	c	(235)
3.	b	(238)
4.	b	(238)
5.	d	(240)
6.	a	(241)
7.	a	(242)
8.	c	(244)
9.	d	(244-245)
10.	c	(247)
11.	a	(249)
12.	d	(251)
13.	b	(254)
14.	d	(256)
15.	c	(257)
16.	d	(260-261)

Matching: Definitions

1.	e	(241)
2.	q	(246)
3.	r	(233)
4.	m	(242)
5.	g	(252)
6.	l	(240)
7.	o	(250)
8.	s	(233)
9.	i	(256)
10.	b	(244)
11.	t	(245)

12.	a	(261)
13.	n	(248)
14.	h	(252)
15.	c	(241)
16.	p	(246)
17.	f	(242)
18.	j	(247)
19.	d	(240)
20.	u	(247)
21.	k	(247)

Completion

1. conservation (240)
2. creativity (234)
3. IQ (236)
4. environmentalists (238)
5. reversible (242)

6. independently (249)
7. obligations; expectations (248)
8. physical (252)
9. testosterone (253)
10. gender identity (252)

Chapter 10

Early Childhood: Integration Into the Human Group

INTRODUCTION

Chapter 9 focused on the expanding competencies of preschool children. In Chapter 10, the early childhood years are once again examined, with primary emphasis on the process of socialization. Chapter 10 is divided into two main topics, including:

- Family Influences. Specific parenting determinants and various child-rearing practices are discussed. Baumrind's four parenting styles and their effects on children's behavior are examined. Research is cited on effective parenting as well as on the sexual abuse of children. The consequences of single-parent families and divorce on parents, as well as on children, are illustrated in detail. The final issue focuses on sibling relationships and their role in the socialization process.

- Peer Relationships. This section discusses children's peer relationships and friendships. The role of children as reinforcing agents and behavioral models for one another is examined, as is the importance of play in the child's personal and social development. The subject of aggression is addressed, with an emphasis on the effects of media violence on the behavior of the nation's youth. Finally, the impact of preschools and Head Start programs on both parents and children is summarized, and a guideline for selecting a preschool is presented.

I. Family Influences

    A. Determinants of Parenting

    B. The Search for Key Child-Rearing Practices

    C. Parenting Styles

    D. Gaining Perspective on Parenting

    E. Single-Parent Families and Children of Divorce

    F. Sibling Relationships

II. Peer Relationships

    A. Peer Relationships and Friendships

    B. Peer Reinforcement and Modeling

    C. Play

    D. Peer Aggression

    E. Bullies and Victims

    F. Television and Video Games

    G. Preschools and Head Start Programs

LEARNING OBJECTIVES

After completing Chapter 10, you should be able to:

1. Define *socialization*, and briefly discuss the role of the family in early human development. (page 266)

2. Identify and briefly discuss Belsky's three major determinants of parental functioning. (267-268)

3. Outline three major dimensions underlying child-rearing and *socialization* practices, and discuss the effects of each of these practices and combination of practices on children's behavior. (268-270)

4. Analyze each of the following parenting styles, and explain how each contributes to the development of socially responsible and independent behavior. (271-273)

   a. *authoritarian*

   b. *authoritative*

   c. *permissive*

   d. *harmonious*

5. Briefly discuss the research of McClelland and White on effective parenting. (274, 276)

6. Summarize the research findings on the *sexual abuse of children*. (Box, 275)

7. Identify some of the difficulties that each of the following people experience (e.g., parent-child contact and communication, changes in behavior of parents as well as children at the time of divorce, and the effects of joint custody):

   a. single parents

   b. divorced parents

   c. children of divorced parents

8. Discuss the impact of no-fault divorce laws on women and children. (Box, 279)

9. Summarize the research findings on sibling relationships, with regard to (282-284)

    a.   birth order

    b.   family size

    c.   *confluence theory*

    d.   *resource dilution hypothesis*

    e.   sex of siblings

10. Characterize the qualitative changes that occur in *peer* friendship patterns. (284-285)

11. Briefly describe the role of children as reinforcing agents and behavioral models for one another. (286)

12. Compare and contrast the different types of **play**, and appreciate the significant contribution that each makes to development. (286-289)

13. Explain the roles played by each of the following factors in the development of *aggressive* behavior: (289-293)

    a.   learning

    b.   competition

    c.   bullying

    d.   television

    e.   video games

14. Assess the effects of preschools and Head Start programs on both parents and children. (294-296)

15. Discuss the advantages and disadvantages of early schooling. (Box, 294-295)

16. Develop criteria for evaluating preschools. (Box, 296)

KEY TERMS

---

aggression (page 289)

authoritarian parenting (271)

authoritative parenting (271)

confluence theory (283)

harmonious parenting (271)

peers (284)

permissive parenting (271)

play (286)

resource dilution hypothesis (283)

scaffolding (272)

sexual abuse of children (275)

socialization (266)

Multiple-Choice

Circle the letter of the response which *best* completes or answers each of the following statements and questions.

Factual Questions:

1. Which of the following is a major determinant of parental functioning?

    a. the number of siblings in the family
    b. the attitudes of the new grandparents
    c. the child's temperament
    d. whether or not the child has a college education

2. Wesley C. Becker, in a review of research on parenting, found that the effectiveness of love-oriented techniques derives from the fact that

    a. they promote children's acceptance of self-responsibility and foster self-control through inner mechanisms of guilt
    b. children reared by love-oriented techniques are subjected to many rules for neatness and proper conduct
    c. permissiveness creates empathy in children and makes them less self-indulgent
    d. they encourage well-controlled, socialized behavior and lead children to perceive the world as dominated by powerful forces

3. Warm but restrictive parenting is believed to lead to _____ according to Maccoby's research on 12-year-old boys.

    a. increased creativity
    b. increased aggression
    c. greater motivation toward schoolwork
    d. increased social competence

4. According to Diana Baumrind's analysis of parental authority, an authoritative parent is best described as one who

    a. stresses obedience and conformity to rules
    b. allows the child to regulate his or her own behavior
    c. prefers forced discipline but allows children to make the rules
    d. provides firm direction but allows the child freedom within limits

5. In a study involving white, middle-class nursery-school children, Diana Baumrind found that the *least* self-reliant, explorative, and self-controlled children were those with parents who were

    a. permissive
    b. authoritative
    c. harmonious
    d. authoritarian

6. _____ supports a child's learning through intervention and tutoring that provide helpful task information attuned to the child's current level of functioning.

    a. Permissive parenting
    b. Scaffolding
    c. Socialization
    d. Harmonious parenting

7. Parents who are the most effective in eliciting compliance from their youngsters and deflecting defiance

    a. assert their power in the form of threats or criticism
    b. employ Baumrind's permissive style of parenting
    c. approach their children's behavior through guidance
    d. employ Baumrind's authoritative style of parenting

8. Studies by McClelland and colleagues have indicated that

    a. breast feeding has an important influence on what people think and do as adults
    b. how parents feel about their children has an important influence on what their children think and do as adults
    c. parents who employ good child-rearing techniques are more likely to raise children who have healthy personalities as adults
    d. parents whose child-rearing techniques are inconsistent are more likely to raise children who become aggressive adults

9. Which statement is the most accurate regarding the sexual abuse of children?

    a. Sexually abused children are usually afraid to tell others about their experiences.
    b. Most research has dealt with the sexual abuse of males.
    c. Most victims of childhood sexual abuse recover from patterns of psychological shame within a few years.
    d. There is a decrease in the number of male victims of childhood sexual abuse.

10. Single-parent families tend to be characterized by

    a. high delinquency rates among the children
    b. poor school adjustment of the children
    c. serious economic problems
    d. male heads of households

11. Which of the following statements is the *least* accurate concerning research on the effects of divorce?

    a. The first year after the divorce was equally stressful for both parents.
    b. Divorced parents communicate less well with their children than parents in intact families.
    c. Poor parenting seems most marked for divorced mothers one year after divorce.
    d. Single-parent families usually experience increased stress and difficulties during the second year of divorce compared with the first year of divorce.

12. According to Weitzman, since the passage of no-fault divorce laws

    a. court-ordered child support payments have been more consistently enforced
    b. women have been deprived of intangible assets acquired during their marriage
    c. more women have been allowed to retain possession of their family homes
    d. long-married women have been more likely to receive alimony as compensation

13. Wallerstein and her associates studied 60 middle-class families undergoing divorce. Follow-up studies on this research indicates that

    a. most young children adapt quite well to divorce if they do not encounter substantial novel sources of personal or family stress
    b. the age of a child at the time of divorce plays an insignificant role in the child's future development
    c. children who are adolescents at the time of divorce are less consciously troubled ten years after the divorce than children who are very young at the time of divorce
    d. five or ten years after a divorce, virtually all children adapt quite well and show no signs of depression or anxiety

14. Researchers find that the quality of children's relationship with _____ is the best predictor of their postdivorce adjustment.

    a. their mother
    b. their siblings
    c. both parents
    d. their father

15. With respect to school achievement, social adjustment, and delinquent behavior, the differences between children from one-parent and two-parent homes of comparable social status are

    a. small
    b. moderate
    c. very large
    d. large

16. Firstborn children are more likely than children occupying other birth-order positions to be

    a. president of the United States
    b. low in measured IQ
    c. underrepresented in the college population
    d. better in athletic ability

17. Studies of peer relationships of children indicate that

    a. 4-year-olds spend about one-third of their time playing with peers
    b. by 18 months of age, social play predominates
    c. spontaneous peer reinforcement decreases with age
    d. a greater amount of social interaction takes place among acquainted toddlers

18. In a classic study, Mildred B. Parten observed the play of children in nursery-school settings. Some children were found to interact with one another, borrowing or lending play materials, following one another about with care, and attempting to influence each other's activity. She termed this type of play

    a. associative
    b. parallel
    c. solitary
    d. cooperative

19. As children grow older, their aggressiveness becomes

    a. more diffuse and less directed
    b. less oriented toward temper tantrums
    c. less retaliatory in response to others' aggression
    d. less verbal and less initiative

20. One of the legitimate arguments against making formal instruction available to 4-year-olds is that

    a. older toddlers have more success during their school careers than their younger counterparts
    b. children must be in the concrete operational stage before they can benefit from formal instruction
    c. formal instruction primarily benefits middle-class children who are cognitively sophisticated
    d. children's intellectual development usually exceeds their social and emotional development

Conceptual Questions:

1. Your clergyman tells you that the best advice in raising children is "Love will show the way." While not disagreeing with your clergyman, you feel that his advice is simplistic because

    a. some parents demonstrate love by accepting and understanding, while others show love through punitive discipline
    b. how consistently you punish your child is a more important consideration than the warmth of your relationship with him
    c. you did not breast-feed your baby, and therefore your relationship with him will not be as warm
    d. some warm parents have well-behaved, dependent, conforming children, while others have more self-reliant and socially assertive children

2. Every time Pedro hits his little sister, his mother says, "That's not nice. Wait until your father gets home, and I will tell him about your misbehavior!" Later that evening, Pedro gets a spanking from his dad. We might reasonably infer that

    a. Pedro will become a submissive and dependent boy
    b. Pedro's mother would have responded more authoritatively if his father were present
    c. Pedro's father is authoritarian and his mother is submissive
    d. Pedro will become an obedient and appropriately aggressive child

3. In your house, the children are all given rules and responsibilities. However, they negotiate with each other for who does what tasks (e.g., dishes), and occasionally one child agrees to do a sibling's share of work if there is a special event (e.g., a school play). On the basis of Baumrind's research, we could predict that

    a. your sons will be very permissive
    b. your sons and daughters will be self-reliant and contented
    c. your daughters will not be very self-reliant
    d. your sons and daughters will be withdrawn and distrustful

4. Theresa is a 5-year-old whose mother rarely asks her to pick up her toys or help around the house and always reasons with Theresa about acceptable standards of behavior. What parenting style does Theresa's mother use?

    a. authoritarian
    b. authoritative
    c. harmonious
    d. permissive

5. Warren is about to become a stepfather. He is most likely to be successful if he

    a. allows the children's mother to continue being the authority figure
    b. is authoritarian and warm
    c. is not demanding but is permissive
    d. is authoritative and warm

6. Samantha's father scolds her when she throws her food on the ground. Samantha's mother, however, does not punish her daughter for this misbehavior. Which of the following conclusions would be the *least* reasonable?

   a. Samantha may become more resistant to future attempts to stop her food-throwing.
   b. Samantha's mother is more likely to scold her daughter's misbehavior if her husband is present.
   c. Samantha will likely develop emotional problems as a result of inconsistent parenting.
   d. Inconsistent punishment will probably fail to put an end to Samantha's food-throwing.

7. Martha's 6-year-old son Cory has thrown his toys all over the house. Which of the following statements should Martha use to best persuade Cory to pick up his mess?

   a. "Could you please pick up the toys now, honey?"
   b. "You better pick up those toys right now, or I will spank you."
   c. "You made a mess, so you'll have to clean it up now."
   d. "You can pick up the mess whenever you want to."

8. Your daughter has recently complained to you about her uncle, who has been fondling her "private parts." You should conclude that

   a. your daughter has an active fantasy life and is probably fabricating this story
   b. her uncle knows better than that, and anyway, child molesters are usually strangers
   c. your daughter is not making this up, and a further conversation is in order
   d. your daughter is too young to be the victim of sexual molestation

9. Karina was sexually abused as a young girl by her stepfather. Which of the following difficulties is Karina likely to experience as a result of this abuse?

   a. learning difficulties
   b. depression
   c. low self-esteem
   d. all of the above

10. Billy's parents have been divorced for one year, and he now lives with his mother. Whenever Billy acts out, his mother is unsuccessful in her efforts to decrease his unwanted behavior. Billy tells her, "You can't make me stop. Only Dad can do the spanking." Psychologists could predict that

    a. as time goes by, Billy's mother will give up on him and make fewer demands for mature behavior
    b. as time goes by, Billy's mother will start ignoring him more and showing him less attention
    c. Billy's school grades will decrease, and he will become more abusive and demanding
    d. Billy would be more obedient for his mother if his father was more psychologically available to him

11. Jillian, the mother of two young children, was recently divorced by her husband. Because of the new divorce laws for women and children in the United States,

    a. Jillian's standard of living is likely to increase 50 percent
    b. Jillian's ex-husband will be required to contribute two-thirds of his income to support his ex-wife and children
    c. valuable assets acquired during Jillian's marriage (e.g., credit, insurance, and future earning power) are likely to follow her ex-husband
    d. most likely, as a result of the divorce, the courts will award Jillian ownership of the family home

137

12. Alice was 3 years old when her parents divorced. Margarita was 14 when her parents divorced. On the basis of research by Wallerstein, we can reasonably infer that

    a. ten years after their parents' divorces, Alice will have adapted better than Margarita to the effects of divorce
    b. because the age of children at the time of divorce matters little, Alice and Margarita should adapt equally well to their parents' divorces
    c. both Alice and Margarita will likely experience no difficulties years after their parents' divorces
    d. Margarita will experience fewer emotional scars than Alice years after their parents' divorces

13. Mary got divorced six months ago after a year of bitter court battles. Which of the following conclusions might be justified?

    a. Her parenting skills will improve because she can devote her time to her children instead of arguing with her husband.
    b. Under these circumstances, a joint custody arrangement might not serve the needs of the children or the parents.
    c. Although her divorce was bitter, her children would have been better off psychologically if she and her husband had stayed together.
    d. She will feel more competent and attractive; and since she will feel better about herself, she will feel better about her children.

14. Firstborn is to intellectual eminence as sibling's sex is to _____

    a. self-esteem
    b. birth order
    c. dethroning
    d. gender role performance

15. You are majoring in early childhood education. At a recent family reunion, you observe some of the children interacting and playing in the family room. Which of the following situations is most likely to be occurring?

    a. Two unacquainted toddler cousins are frolicking with their arms embraced, while two cousins who live next door to one another are fighting.
    b. Two 4-year-old cousins are encouraged by their uncles to play with each other instead of hanging around the adults.
    c. The boy and girl preschoolers are in a large group holding hands and playing a game of kickball.
    d. Two-year-old children are smiling at one another, babbling, and exchanging toys.

16. At nursery school, Jessie is playing next to some other children. All are playing with blocks, but Jessie is not interacting with the others. Jessie is

    a. engaging in onlooker play
    b. more likely to be middle-class than lower-class
    c. more likely to be female than male
    d. engaging in object-centered contacts

17. Your husband is always telling the children, "Quit playing around and do something useful with your time." Your best advice to your husband might be that

    a. although play is a poor use of time, children need play to expend excess energy
    b. although most play is a waste of time, middle-class children generally use play time efficiently
    c. telling this to the children will decrease their enjoyment of play later in life
    d. play is the child's work; it has vital social, cognitive, and physical benefits for the child

18. You are a nursery-school teacher and are having trouble with a number of youngsters who are very aggressive. Psychologists might recommend that you

    a. let the bullies and their victims role-play, switching roles
    b. have all the children play games like ring toss in order to win a limited number of prizes
    c. demonstrate methods of verbal instead of physical retaliation
    d. segregate the children so that the boys play with boys, and the girls play with the girls

19. You read, with alarm, that children observe 18,000 murders on the television screen by the time they finish high school. On the basis of the research, which of the following would be a reasonable conclusion?

    a. There is a hidden socializer in your home that you may not be aware of--the television set.
    b. You should not be concerned about your child's television viewing because television is a vicarious medium.
    c. You should not be concerned because the "violence" level of children's shows is carefully controlled.
    d. You should allow your child to watch only those violent shows that have prosocial endings.

20. Both you and your wife are college educated and work outside the home in professional situations. Your neighbor who runs an academic preschool program encourages you to enroll your 4-year-old so that your child can get a jump on the other neighborhood children and learn to read and write. Your best reply to your neighbor might be,

    a. "I feel that early ripe means early rot."
    b. "All work and no play makes Johnny a dull boy."
    c. "You're right; a mind is a dangerous thing to waste."
    d. "Anything you can do, I can do better."

Matching: Definitions

Match the letter of the term in the box with its definition.

```
a.   solitary                    i.   vicarious conditioning
b.   permissive                  j.   harmonious
c.   socialization               k.   video games
d.   authoritative               l.   aggression
e.   confluence                  m.   scaffolding
f.   authoritarian               n.   resource dilution
g.   peers                       o.   associative
h.   play
```

1. type of play that occurs when children play by themselves _____

2. parenting style that provides a nonpunitive accepting environment in which children regulate their own behavior _____

3. parenting style that provides firm direction for a child's activities within reasonable limits _____

4. theory which states that older children experience a richer intellectual environment than their younger siblings _____

5. behavior that is socially defined as injurious or destructive _____

6. parenting style that emphasizes traditional, absolute values and standards of conduct as well as obedience and forceful discipline _____

7. the process of transforming children into functioning members of society _____

8. activities which are ends in themselves _____

9. individuals of approximately the same age _____

10. parenting style in which parents exercise little control, cultivate egalitarian relationships, and emphasize humane values _____

11. type of observational learning in which children acquire behavior by imaginatively participating in an experiment _____

12. type of play that occurs when children interact with one another, and no division of labor or integration of activity takes place _____

13. supports a child's learning through intervention and tutoring that provide helpful task information attuned to the child's current level of functioning _____

14. theory which states that in large families resources get spread thin, to the detriment of all the offspring _____

15. allow youngsters little opportunity to make decisions for themselves, to fashion their own fantasies, and in turn to construct their own problem resolutions _____

Completion

Supply the term or terms needed to complete each of the following statments.

1. Parents who display _____ toward their children tend to employ frequent explanations, words of encouragement, and praise when disciplining them.

2. According to Baumrind, parenting that is _____ tends to produce children who are the least self-reliant, explorative, and self-controlled.

3. _____ _____ of children is commonly defined as forced, tricked, or coerced sexual behavior between a child and an older person.

4. _____ divorce laws were designed by their framers to treat men and women as equals.

5. Divorced parents tend to make fewer maturity demands of their children, to communicate less with them, to be less affectionate with their children, and to show marked _____ in disciplining them.

6. According to the predictions of _____ theory, children with a greater number of older siblings will achieve lower intellectual levels.

7. _____ peers have more positive interactions with each other and engage in less conflict than _____ peers.

8. According to Parten, a developmental progression from parallel to associative to _____ play occurs in most children.

9. Children who are _____ tend to have a perceptual basis that leads them to see and retaliate against threats where none exist.

10. Recent studies of children who had participated in _____ _____ preschool programs show that they have performed as well or better than their peers in regular school.

Multiple-Choice

Factual			Conceptual		
1.	c	(268)	1.	d	(269)
2.	a	(269)	2.	b	(271)
3.	c	(269)	3.	b	(271)
4.	d	(271)	4.	d	(271)
5.	a	(271)	5.	d	(271)
6.	b	(272)	6.	c	(271)
7.	d	(271-272)	7.	c	(272-273)
8.	b	(274)	8.	c	(275)
9.	a	(275)	9.	d	(275)
10.	c	(277)	10.	c	(278-279)
11.	d	(278)	11.	c	(279)
12.	b	(279)	12.	a	(280)
13.	a	(280)	13.	b	(281)
14.	c	(281)	14.	d	(282-284)
15.	a	(281)	15.	d	(284-285)
16.	a	(282)	16.	c	(288)
17.	d	(284-285)	17.	d	(288-289)
18.	a	(288)	18.	a	(291)
19.	b	(289)	19.	a	(292-293)
20.	a	(295)	20.	b	(294-295)

Matching: Definitions

1.	a	(287)	9.	g	(284)
2.	b	(271)	10.	j	(271)
3.	d	(271)	11.	i	(292)
4.	e	(283)	12.	o	(288)
5.	l	(289)	13.	m	(272)
6.	f	(271)	14.	n	(283)
7.	c	(266)	15.	k	(293)
8.	h	(286)			

Completion

1.	warmth (269)	6.	confluence (283)
2.	permissive (271)	7.	acquainted; unacquainted (284-285)
3.	sexual abuse (275)	8.	cooperative (288)
4.	no-fault (279)	9.	bullies (291)
5.	inconsistency (278)	10.	Head Start (295)

Chapter 11

Later Childhood:
Advances in Cognitive and Moral Development

INTRODUCTION

Chapter 11 presents an overview of the advances in cognitive and moral development that occur during later childhood. These changes are discussed within the context of the developing child during his or her elementary school years--ages 7 to 12. Several topics are covered, including:

- Emerging Cognitive Abilities. This is a discussion of children's emerging cognitive capabilities, which permit them to organize and process environmental stimuli. Piaget's stage of concrete operations is presented as characterizing the elementary school child. The various types of learning styles which children rely on in solving problems as well as how they acquire socialization skills are also reviewed.

- Emotional Functioning. The text describes the relationship between children's intellectual and emotional development. Children's emotional functioning is guided and shaped by the cognitive interpretation they assign to specific experiences. Also discussed within this context is the impact that fear and stress have on children's functioning.

- Moral Development. The text presents the major theories used to understand children's moral development: psychoanalytic, cognitive learning, and cognitive-developmental theories. Also included is a review of the inconsistency of children's moral behavior, as well as the various personal and situational factors most closely associated with moral behavior.

- Prosocial Behavior. The chapter concludes with an overview of the ways in which children acquire prosocial behaviors.

CHAPTER OUTLINE

I. Cognitive Sophistication

    A. Period of Concrete Operations

    B. Cognitive Style

    C. Person Perception

II. Understanding Emotion and Dealing with Fear

III. Moral Development

    A. Psychoanalytic Theory

    B. Cognitive Learning Theory

    C. Cognitive-Developmental Theory

    D. Moral Behavior: Consistency or Inconsistency?

    E. Correlates of Moral Conduct

IV. Prosocial Behavior

LEARNING OBJECTIVES

After completing Chapter 11, you should be able to:

1. Describe the development of *executive strategies* during the *period of concrete operations* by focusing on the following abilities: (pages 302-304)

    a. *conservation*

    b. horizontal decalage

    c. decentration

    d. attention to transformations

    e. reversibility

2. Identify the characteristics of *psychomotor skills* and the phases in the learning of these skills. (Box, 303)

3. Formulate a position on Piaget's "American question" by evaluating cross-cultural research on *conservation* ability. (306)

4. Define *cognitive style*. (306)

5. Summarize the research cited in the text on differences in children's *cognitive styles* (impulsive versus reflective, field independence versus field dependence). (306-307)

6. Identify the changes that occur in children's person-perception abilities. (308-309)

7. Describe the qualitative, age-related differences in regard to children's descriptive statements about other people. (308)

8. Discuss situations which children commonly *fear* at various ages. (309-310)

9. List several techniques that psychologists have found useful for helping children deal with *fear*. (310)

10. Define *stress* and *coping*. (312)

11. Discuss the factors which influence how children experience and deal with *stress*. (313)

12. Compare and contrast the following theories of *moral development*: (311-314)

    a. psychoanalytic theory

    b. cognitive learning theory

13. Summarize the viewpoints presented by the following cognitive-developmental theorists: (315-320)

    a. Jean Piaget

    b. Lawrence Kohlberg

    c. Elliot Turiel

    d. William Damon

14. Identify examples of children's behavior that are associated with each of the different stages or levels in Piaget's, Kohlberg's, and Damon's theories of moral development. (315-320)

15. Explain the significance of the Hartshorne and May research on the "consistency of moral behavior." (320-321)

16. Discuss what Robert Coles has found in regard to children's vulnerabilities and strengths. (322-323)

17. Assess the role played by the following in influencing moral conduct: (321-323)

   a. intelligence

   b. age

   c. sex

   d. group norms

   e. motivational factors

18. Discuss the factors that influence children's prosocial behaviors and provide some typical examples of such behaviors. (324-325)

KEY TERMS

---

altruism (page 324)

autonomous morality (315)

cognitive style (306)

conservation (303)

coping (312)

executive strategy (302)

fear (309)

heteronomous morality (315)

internalize (312)

locus of control (312)

moral development (311)

period of concrete operations (302)

147

phobia (309)

prosocial behavior (altruism) (324)

psychomotor skill (303)

stress (312)

SELF-TESTS - CHAPTER 11

Multiple-Choice

Circle the letter of the response which *best* completes or answers each of the following statements and questions.

Factual Questions:

1. During the autonomous phase of psychomotor skill acquisition, children

    a. identify the separate parts of a skill
    b. fit individual skills into a sequential order
    c. require little conscious attention or regulation in performing an activity
    d. eliminate inappropriate responses and become more goal directed

2. Children become capable of decentering when liquid is poured from one container into another during which of the following stages?

    a. formal operational
    b. sensorimotor
    c. preoperational
    d. concrete operational

3. According to Jean Piaget, children in the concrete operational stage are able to recognize that an original state can be regained. They recognize that water poured from one container into another can be poured back into the original container. This is called

    a. decentering
    b. transformations
    c. reversibility
    d. horizontal decalage

4. During the period of concrete operations, children acquire and develop specific skills in a sequential manner. This type of development, with each skill dependent on the acquisition of earlier skills, is termed

    a. conservation
    b. transformations
    c. executive development
    d. horizontal decalage

5. Cross-cultural research on the acquisition of conservation skills indicates that

   a. these skills may not occur in the invariant sequence postulated by Piaget
   b. non-Western children are ahead of their Western counterparts in developing conservation skills
   c. conservation skills develop only in children who achieve formal mental operations
   d. Western children are more likely than children from nonindustrialized cultures to attribute magical action to the experimenter

6. The stable preferences that individuals exhibit in organizing and categorizing their perception *best* defines

   a. conservation
   b. altruism
   c. cognitive style
   d. reversibility

7. Research suggests that field-dependent college students are most likely to major in

   a. natural sciences
   b. mathematics
   c. engineering
   d. humanities

8. Children under 8 years of age tend to describe other people in

   a. terms of external, readily available attributes
   b. terms of unchanging qualities and inner dispositions
   c. specific, precise, and concrete ways
   d. a straightforward, nonstereotypic way

9. Which of the following would be the best strategy to follow in helping a child overcome fears?

   a. Create an accepting situation in which the child feels at ease in sharing his fears with you.
   b. Shame the child in front of others because of his or her fear.
   c. Goad the child into trying things he or she is not ready for.
   d. Completely ignore the child's fears or attempt to make fun of the fear.

10. Individual perception of who or what is responsible for the outcome of events and behavior in one's life is known as

    a. cognitive style
    b. person perception
    c. locus of control
    d. innate disposition

11. Jean Piaget indicated that heteronomous morality is characterized by which of the following?

    a. mutual respect
    b. morality of cooperation
    c. an egalitarian attitude
    d. conception of moral rules as absolute

12. At which level of Kohlberg's theory of moral development are individuals capable of differentiating between themselves and the rules and expectations of others?

    a. conventional level
    b. postconventional level
    c. heteronomous morality level
    d. autonomous morality level

13. According to Damon, the most morally advanced children reason about distributive justice in the following way:

    a. they assert their choices and do not attempt to justify them
    b. they justify their choices on the basis of external, observable traits
    c. they balance conflicting claims derived from notions of equality
    d. they evolve notions in which fairness is equated with strict equity

14. Research undertaken by Hugh Hartshorne and Mark A. May involving the moral behavior of some 11,000 children revealed that children's moral behavior was

    a. very highly consistent
    b. highly consistent
    c. moderately consistent
    d. quite inconsistent

15. Research reveals that age, sex, and IQ variables are _____ correlated with moral behavior.

    a. very highly
    b. highly
    c. moderately
    d. minimally

16. In a follow-up study in which respondents completed a psychological measure of empathy, researchers found that the single most powerful predictor of empathy in adulthood was

    a. how much time the subjects' mothers had spent with them when they were children
    b. how much time the subjects' fathers had spent with them
    c. the parenting style of the respondents' fathers
    d. the parenting style of the respondents' mothers

Conceptual Questions:

1. In teaching a child how to play baseball, we may first teach him how to swing a bat at the ball placed on a stationary pedestal (T-ball). By doing this, the child masters the arm and leg movements involved in the swing before he coordinates these with eye movements. A child in the T-ball stage of baseball is in the

    a. autonomous phase of psychomotor learning
    b. associative phase of cognitive learning
    c. cognitive phase of psychomotor learning
    d. executive strategy phase of motor learning

2. Kara observes that her son has developed the various conservation skills in a sequential manner. According to Piaget, this type of successive skill development is referred to as

    a. locus of control
    b. conventional development
    c. horizontal decalage
    d. decentration

3. Margaret watches her mother pour water from a short, wide container into a long, narrow flower vase. She realizes that the initial state of the water can be regained by pouring it back into the original container. Piaget would say Margaret has attained which of the following?

   a. reversibility
   b. transformation
   c. decentration
   d. internal locus of control

4. A 5-year-old child acquires conservation of number by watching an older child being rewarded for giving correct answers in a similar task. Piaget would say that this 5-year-old child's

   a. behavior was a positive response to an American question
   b. behavior was an example of horizontal decalage
   c. upbringing was in a third-world country
   d. behavior indicated a reflective cognitive style

5. Ten-year-old Neil is very good at a game called "word search" in which he must find a word hidden among a bunch of randomly arranged letters. Psychologists would predict that Neil

   a. probably has a reflective cognitive style
   b. would probably benefit from a lecture-discussion classroom style
   c. would do well in an analytic discipline like mathematics
   d. will display verbal flexibility and originality

6. After her parents were divorced, 10-year-old Jennifer remarked, "Even though Mom and Dad act like they like each other, I don't really think they do." This statement indicates

   a. the use of simple moralistic social categories
   b. insightful person-perception ability
   c. an inability to activate social stereotypes
   d. that Jennifer has internalized moral prohibitions

7. Julie's parents are concerned because although she is 10 years old, she is still afraid of water and refuses to learn to swim. The best suggestion you could provide her parents would be to

   a. tell Julie you will buy her a new bike if she learns to swim
   b. allow Julie to observe other children enjoying themselves in the water at the beach
   c. tell Julie that she is silly for being afraid of the water; it can't really hurt her
   d. firmly take Julie into the water by saying, "I won't hurt you or let your head go under; you'll be fine"

8. Matthew and Sarah are concerned about how their child will be able to deal with the stress resulting from their divorce. Research has demonstrated that they should attend to each of the following factors in helping their child cope *except* for

   a. the dispositional and temperamental tendencies of their child
   b. their child's developmental level
   c. their child's ability to ignore the situation
   d. situation-specific factors (e.g., parental reaction)

151

9. In which of the following situations should parents be the most concerned about their child's coping abilities?

   a. Four-year-old John attends his grandfather's funeral and states, "Can I play with Grandpa if he wakes up from his sleep?"
   b. Michelle, the leading gymnast in the elementary school, loses a competition to a new student in her class.
   c. Ryan, a junior high student, does not seem particularly distressed when his mother informs him that she is planning to divorce his father.
   d. Six-year-old Suzanne is reluctant to go in the hospital for leukemia treatments because her mother seems anxious about the side effects.

10. A young man in your neighborhood bullies the young children and terrorizes the neighborhood pets. Which of the following conclusions would be the most reasonable if you were a cognitive learning theorist?

    a. The young man's superego is overdeveloped.
    b. The young man will probably act this way wherever he lives.
    c. The young man has not developed a tabula rasa.
    d. The young man behaves in this manner because his actions usually lead to positive consequences.

11. A sixth grade student displays a willingness to cooperate with his fellow classmates. Piaget would characterize this child's egalitarian relationship with his peers as being representative of which stage of moral development?

    a. autonomous morality
    b. heteronomous morality
    c. conventional level
    d. preconventional level

12. According to Kohlberg, which of the following children is demonstrating the highest level of moral development?

    a. Billy, who treats his sister nicely when Mom and Dad are around so that he won't get yelled at
    b. Jimmy, who tells his mother that he got good grades because he knows his mother will like hearing this
    c. Don, who cleans the blackboards for his teacher so he might get a better grade
    d. Mary Beth, who pickets in front of an abortion clinic even though the law allows abortions to be performed

13. Donny tells his older brother, "I won't tell Mom that you were kissing your girlfriend in the living room if you don't tell her I was eating cheese and crackers in the bedroom." Donny is demonstrating

    a. autonomous morality
    b. concrete operational reasoning
    c. conventional morality
    d. an instrumental orientation to moral reasoning

14. Sam is an Eagle Scout and has been praised for his honesty and trustworthiness. However, yesterday Sam got caught stealing from the grocery store. On the basis of psychological research, the most logical conclusion we can reach is that

    a. Sam's stealing is motivated by a universal ethical orientation
    b. Sam's honesty is specific to particular situations and not a trait of his character
    c. Sam is morally immature because his moral attitudes are inconsistent with his behavior
    d. if Sam were more intelligent, he would have avoided being caught

15. Which of the following statements is *not* true regarding Robert Coles' detailed analysis of children's lives?

    a. Critics contend that his research is insufficient and unscientific.
    b. Children wonder about the same types of questions that we all have wondered about.
    c. His approach has been criticized because he is too detached.
    d. He concentrated on studying the lives of "normal" children.

Matching: Definitions

Match the letter of the term in the box with its definition.

a.	cognitive style	k.	prosocial behavior
b.	heteronomous morality	l.	phobia
c.	metacognition	m.	tabula rasa
d.	coping	n.	internalize
e.	moral development	o.	autonomous morality
f.	horizontal decalage	p.	period of concrete operations
g.	fear	q.	empathy
h.	conservation	r.	psychomotor skill
i.	desensitization	s.	altruism
j.	executive strategy	t.	stress

1. the responses people make in order to master, tolerate, or reduce stress _____

2. feelings of emotional arousal that lead a child to take the perspective of another _____

3. a process involving the recognition of and response to threat or danger _____

4. an unpleasant emotion aroused by impending danger, pain, or misfortune _____

5. sequential skill development, with each skill dependent on the acquisition of earlier skills _____

6. Piaget's first stage of moral development in which rigid and absolute judgments are derived from moral rules laid down by persons in authority _____

7. the concept that the quantity or amount of something stays the same regardless of any changes in its shape or position _____

8. conveys the notion that children are morally neutral _____

9. an excessive, persistent, and unadaptive fear response (usually to ill-defined stimuli) _____

10. a process by which individuals are led by degrees into active contact with and participation in situations that they fear _____

11. behavior that is carried out to benefit another without expectation of an external reward _____

12. children's growing awareness and understanding of their own mental processes _____

13. process by which children adopt principles that lead them to evaluate given behaviors as "right" or "wrong" and to govern their actions accordingly _____

153

14. a method for selecting, sequencing, evaluating, revising, and monitoring a problem-solving plan and behavior _____

15. a stage of development in Piaget's theory in which classifying and ordering activities begin to occur internally as mental abilities _____

16. stable preferences that individuals exhibit in organizing and categorizing perceptions _____

17. ways of responding to others that include sympathetic, cooperative, helpful, rescuing, comforting, and giving acts _____

18. what children do when they incorporate within their personality the prohibitions of their parents _____

19. a mind-guided muscular movement and its coordination _____

20. Piaget's second stage of moral development in which moral judgments derive from an egalitarian perspective based on mutual respect and cooperation _____

Completion

Supply the term or terms needed to complete each of the following statements.

1. During the _____ phase in the learning of motor skills, people no longer need to "think about what to do next."

2. Concrete operational children are able to attend to _____, for example, the gradual shift in the height or width of the fluid in a container as it is poured.

3. Cross-cultural research suggests that there is a developmental lag in the acquisition of _____ among children in non-Western, nonindustrialized cultures.

4. _____ people tend to analyze the elements of a scene; they focus on items as being separate from their backgrounds.

5. When people perceive the outcome of an action as the result of luck, change, fate, or powerful others, they believe in _____ control.

6. Piaget's stage of autonomous morality is also called the morality of _____ stage.

7. Kohlberg's _____ level of moral development is characteristic of most children under 9 years of age.

8. Research has shown that children's moral behavior is varied to suit the _____.

9. _____ implies that behavior is carried out to benefit the other person without the expectation of an external reward.

10. Research reveals that adults' _____ speak louder than their words in regard to influencing their childrens' development of prosocial behaviors.

Multiple-Choice

Factual

1.	c	(303)
2.	d	(303-304)
3.	c	(304)
4.	d	(304)
5.	a	(306)
6.	c	(306)
7.	d	(307)
8.	a	(308)
9.	a	(310)
10.	c	(312)
11.	d	(315)
12.	b	(316)
13.	c	(320)
14.	d	(320)
15.	d	(321-322)
16.	b	(325)

Conceptual

1.	c	(303)
2.	c	(304)
3.	a	(304)
4.	a	(306)
5.	a	(307)
6.	b	(308-309)
7.	b	(310)
8.	c	(313)
9.	d	(313)
10.	d	(313-314)
11.	a	(315)
12.	d	(317)
13.	d	(317)
14.	b	(322-323)
15.	c	(322-323)

Matching: Definitions

1.	d	(312)		11.	s	(324)
2.	q	(324)		12.	c	(302)
3.	t	(312)		13.	e	(311)
4.	g	(309)		14.	j	(302)
5.	f	(304)		15.	p	(302)
6.	b	(315)		16.	a	(306)
7.	h	(303)		17.	k	(324)
8.	m	(311)		18.	n	(312)
9.	l	(309)		19.	r	(303)
10.	i	(310)		20.	o	(315)

Completion

1.	final or autonomous (303)		6.	cooperation (315)
2.	transformations (304)		7.	preconventional (316)
3.	conservation (306)		8.	situation (320-321)
4.	field-independent (307)		9.	altruism (324)
5.	external (312-313)		10.	actions (325)

Chapter 12

Later Childhood: The Broadening Social Environment

INTRODUCTION

Chapter 12 describes a variety of ways in which the school environment influences children's development. More specifically, this chapter illustrates how the school setting affects the development of a child's personality, cognitive capabilities, interpersonal skills, and school behavior. Several important issues are discussed, including:

- The Vital Role of Peer Relationships. Several areas are covered within this context: functions of the peer group, influence of same-sex relationships, the consequences of social acceptance and rejection, development of children's self-esteem and the factors that contribute to a healthy self-concept, the processes by which children conform to a peer group, and children's awareness of and experience with members of different social and ethnic groups.

- The Nature and Purpose of Schools. The areas examined include the direct functions of the school, characteristics of effective schools, identification and placement of children with learning disabilities, factors associated with improving student motivation, and the close relationship between school performance and socioeconomic status.

I. The World of Peer Relationships

   A. Developmental Functions of Peer Groups

   B. Same-Sex Groupings

   C. Social Acceptance and Rejection

   D. Self-Esteem

   E. Conformity to the Peer Group

   F. Racial Awareness and Prejudice

II. The World of School

   A. Developmental Functions of Schools

   B. The Effectiveness of Schools

   C. Motivating Students

   D. School Performance and Social Class

LEARNING OBJECTIVES

After completing Chapter 12, you should be able to:

1. Summarize the significant functions of children's peer relationships and peer groups. (pages 330-331)

2. Describe the changes that occur as children pass from a *sex cleavage* orientation to a heterosexual orientation to social behavior. (331-334)

3. Explain the factors which, according to developmental psychologist Maccoby, contribute to gender segregation among youngsters. (332)

157

4. List the various stages which, according to Broderick and Rowe, children typically pass through during heterosexual development. (333-334)

5. Specify the relationship between the following terms: *group*, *values*, and *sociometry*. (334)

6. Recall what the research has found in regard to the physical qualities which make children appealing or unappealing in the eyes of their peers. (334-335)

7. Discuss the behavioral characteristics which seem to be related to children's social acceptance by their peers. (335-336)

8. Illustrate the effects of children's social maturity on their social desirability by their peers. (336)

9. Explain the advantages and disadvantages of using drugs to help hyperactive children. (337)

10. Discuss the meaning, origins, and factors associated with the development of self-esteem in children. (336-339)

11. Explain how the following phenomena are assessed in children: (339-342)

    a. conformity and its influencing factors

    b. racial awareness and *prejudice*, and their influencing factors

12. List the developmental functions that *schools* serve; on the basis of research found in the text, assess how effective schools are in meeting these functions (e.g., whether *mainstreaming* works for the *learning-disabled*). (342-345)

13. Discuss the following factors as they relate to student *motivation*: (345-347)

    a. *intrinsic* and *extrinsic motivation*

    b. attributions of causality

    c. locus of control

14. Compare the academic achievement of Japanese, Taiwanese, and American children; and offer explanations for your findings. (348-350)

15. Describe the relationship among social class, *educational self-fulfilling prophecies*, and school performance. (347-351)

educational self-fulfilling prophecies (page 351)

extrinsic motivation (346)

group (334)

intrinsic motivation (346)

learning disability (344)

mainstreaming (344)

motivation (346)

prejudice (341)

school (342)

sex cleavage (331)

sociometry (334)

value (334)

SELF-TESTS - CHAPTER 12

Multiple-Choice

Circle the letter of the response which *best* completes or answers each of the following statements and questions.

Factual Questions:

1. Which of the following functions is performed by peer groups?

    a. Peer groups give children experience with relationships in which they are on an equal footing with others.
    b. Peer groups help teach children that they must subordinate group goals to their own interests.
    c. Peer groups provide an arena in which children cannot be independent of adult control.
    d. Peer groups provide children with marginal stress.

2. The tendency for elementary school boys to associate with boys and elementary school girls to associate with girls is called

   a. gender consciousness
   b. peer groups
   c. sex cleavage
   d. gender experimentation

3. According to Eleanor Maccoby's findings on gender segregation, which of the following statements is *not* true?

   a. Children systematically frustrate adult efforts to diffuse their preferences for interacting with same-sex peers.
   b. Children segregate themselves into same-sex groups because they find play partners of the same sex more compatible.
   c. Girls find it difficult to establish reciprocity in their relationships with boys.
   d. Boys in their own groups are more likely to engage in "collaborative speech acts."

4. Broderick and Rowe studied the sexual interests of school children between 10 and 12 years of age and found that preadolescents

   a. exhibit the developing roots of adult heterosexual interests and activities
   b. exhibit no interest in members of the opposite sex
   c. reject all members of the opposite sex until they reach adolescence
   d. have predominantly opposite-sex friendships

5. Members of a group

   a. are bound together in relatively stable patterns of social interaction
   b. are attracted to each other as a result of sociometry
   c. share a temporary feeling of unity until a sociogram is developed
   d. join together to preserve a sense of independence

6. Sociometry is a

   a. meter instrument for counting social interactions
   b. measure used for assessing patterns of attraction, rejection, or indifference
   c. direct measure of the desirability of one's name
   d. measure of physical attractiveness

7. Popular children tend to be described by their peers as

   a. noisy, rebellious, and peppy
   b. alert, self-assured, and helpful
   c. confident, aggressive, and attention-seeking
   d. active, egocentric, and strong

8. In many cases parents, teachers, and peers report a significant improvement in the classroom behavior of hyperactive children as a result of administering

   a. amphetamines
   b. antidepressants
   c. tranquilizers
   d. megavitamins

9. Children with high self-esteem are most likely to have parents who

   a. have low self-esteem
   b. set vaguely defined limits
   c. show respect for their children's opinions
   d. abuse their children

10. Studies reveal that in conformity experiments employing the highly ambiguous autokinetic effect, conformity tends to _____ with increasing age.

    a. remain the same
    b. increase
    c. decline
    d. dissipate

11. During the preschool period, white children

    a. show consistent prejudice toward black children
    b. attend less to differences among blacks than they do among whites
    c. use hair and eye characteristics to discriminate between blacks and whites
    d. perceive and think about blacks in stereotyped ways

12. In an effort to overcome the education deficits and difficulties of children with learning disabilities, the schools have used _____ to integrate these children into regular school programs.

    a. tutorials
    b. self-fulfilling prophecies
    c. mainstreaming
    d. guided parenting

13. Which of the following is the most closely associated with intrinsic motivation?

    a. praise
    b. an increase in allowance
    c. increased effort for no external reason
    d. achievement of honor roll for money

14. Studies have shown that the higher the social class of children's families,

    a. the lower the number of elective offices they hold in school life
    b. the lower their participation in extracurricular activities
    c. the higher their rate of truancy and suspensions from school
    d. the greater number of formal grades the children complete

15. Compared with children in Taiwan and Japan, American children

    a. are treated more as autonomous learners
    b. spend more time on academic activities
    c. are encouraged less by their parents to do homework
    d. are socialized to believe in the value of native ability

Conceptual Questions:

1. You are very distressed because your third-grade son has formed a club with his buddies. They all wear the same color shirts, meet in a tree house in the backyard, and talk pig latin. They also have a sign on the treehouse that says "NO GIRLS ALLOWED." Your son's behavior

    a. will become less sexist when he develops sex cleavage
    b. is helping to enhance his self-esteem
    c. is a form of delinquent nonconformity
    d. is abnormal because he should at least be chasing and teasing his sister

2. You and your husband have a discussion and decide that it is about time to sit down and have a talk with your 12-year-old son about "the birds and the bees." Psychologists would tell you to be the most surprised if your son told you that he

    a. disliked all girls and displayed no sexual interests
    b. expected to get married some day
    c. had a girlfriend
    d. would like to go out on a date

3. The personnel review committee at your office decides to reject a job applicant who came in for a job interview wearing a pizza-stained bowling shirt. The group's action indicates the

    a. conservative-shift phenomenon
    b. Asch conformity effect
    c. similarity of self- and reflected appraisal
    d. presence of shared values

4. A sixth-grade student is described by his peers as noisy, attention-seeking, and arrogant. According to behavioral research, he would most likely be labeled as which of the following?

    a. social isolate
    b. mesomorph
    c. hyperactive
    d. introverted

5. Which of the following assessment findings should be the *least* valued in determining whether or not a child exhibits ADHD?

    a. short attention span
    b. inability to sit still
    c. high frequency of discipline problems
    d. teacher's report of his poor academic performance

6. Your child has trouble paying attention in class, has discipline problems, and tends to throw temper tantrums. However, the behavior improves tremendously when your doctor prescribes amphetamines. Your child's diagnosis would probably be

    a. sociopathic behavior
    b. early childhood autism
    c. attention-deficit hyperactivity disorder
    d. ego-resiliency disorder

163

7. Sociologists like Cooley and Mead would maintain that other people serve the role of a(n) _____ in affecting self-esteem.

    a. magnifying glass
    b. mirror
    c. microscope
    d. amplifier

8. Your neighbor's 8-year-old son is always putting himself down when he speaks and relies on other children to tell him what to do. On the basis of the research on self-esteem, you might conclude that his

    a. father does not allow him to engage in give-and-take discussions concerning how the house chores are decided
    b. father treats him in a matter-of-fact, neutral manner while praising good behavior
    c. mother is a loving parent who allows him to break rules occasionally
    d. father lets him do everything for himself and then praises him liberally

9. You want your daughter to be a leader instead of a follower. You challenge her by saying, "If all your friends jumped into the lake in the middle of winter, would you jump in, too?" Her answer would more likely be "Yes" if

    a. she knew the temperature of the water
    b. she didn't know the temperature of the water
    c. her teacher was taking a swim, too
    d. some children jumped in and others did not

10. Which implementation would be *most* effective in promoting a sense of interracial friendliness in a school district?

    a. an interracial basketball league for junior high students
    b. an interracial chess club for students in middle school
    c. an interracial soccer league for elementary school students
    d. an interracial learning team made up of preschool children

11. Which of the following is *most* likely to be characteristic of a youngster who has been classified as having a learning disability?

    a. socially impaired
    b. emotionally disabled
    c. below average in intelligence
    d. difficulty with reading

12. Roberta believes that, regardless of her diligent work in her seventh grade science class, her grade will ultimately be determined by luck. According to research on the attributions of causality, she is most likely to have

    a. an external locus of control
    b. a need to be intrinsically motivated
    c. an internal locus of control
    d. a need to be extrinsically motivated

164

13. Takeshi, a senior high school student of Asian descent, has recently been awarded a prestigious educational achievement award. Which of the following factors has been *least* significant in contributing to his success?

    a. his family's strong ethical code
    b. his inability to move upward in the entertainment business
    c. his parents' constant praise of his advanced intellectual ability
    d. his parents' own well-educated background

14. Maria's family is considered to be in the upper class in socioeconomic status (SES). Psychologists would be the *least* likely to attribute the positive relationship between Maria's high educational achievement and her SES to which of the following?

    a. bias of teachers toward middle- and upper-class youngsters
    b. subcultural differences in regard to their expectations about academics
    c. educational self-fulfilling prophecies
    d. the low value assigned to education by parents of middle- and upper-class children

15. Your son is an average "C" student, and your daughter is an "A" student. Your daughter's teacher gets your son for a class and thinks that he is an "A" student, too. Your son's grades improve. This demonstrates

    a. sexual cleavage
    b. the middle-class bias effect
    c. the effects of excellent teaching
    d. an educational self-fulfilling prophecy

Matching: Definitions

Match the letter of the term in the box with its definition.

a. prejudice	i. mainstreaming
b. sex cleavage	j. sociometry
c. intrinsic motivation	k. school
d. group	l. learning disability
e. motivation	m. mesomorph
f. educational self-fulfilling prophecy	n. sociogram
g. value	o. reflected appraisals
h. extrinsic motivation	p. autokinetic effect

1. the tendency for elementary school boys to associate with boys and elementary school girls to associate with girls _____

2. an umbrella concept referring to difficulty with school-related material despite the appearance of normal intelligence and lack of a demonstrable physical, emotional, or social impairment _____

3. depicts the patterns of choice existing among members of a group at a given time _____

4. an objective method for assessing patterns of attraction, rejection, or indifference among group members _____

5. children fail to learn because those who are charged with teaching them do not believe that they will learn, do not expect that they can learn, and do not act toward them in ways that help them to learn _____

6. refers to the apparent movement of a fixed spot of light in a darkened room _____

7. two or more people who share a feeling of unity and are bound together in relatively stable patterns of social interaction _____

8. those inner states and processes that prompt, direct, and sustain activity _____

9. an agency that provides formal, conscious, and systematic training _____

10. an activity that is undertaken for its own sake _____

11. a criterion that people use in deciding the relative merit and desirability of things (themselves, other people, objects, events, ideas, facts, and feelings) _____

12. activity that is undertaken for some purpose other than its own sake _____

13. suggests that our self-conceptions emerge from social interaction with others and that our self-conceptions in turn influence and guide our behavior _____

14. individuals with an athletic, muscular, and broad-shouldered build _____

15. the integration of students with special needs within the regular programs of the school _____

16. a system of negative conceptions, feelings, and action orientations regarding the members of a particular religious, racial, or nationality group _____

Completion

Supply the term or terms needed to complete each of the following statements.

1. The peer group is the only social institution in which the position of children is not _____.

2. Sigmund Freud referred to the elementary school years as the _____ period.

3. Before age 6, children's conceptions of physical attractiveness tend to be highly _____.

4. _____ children, while interested in others, do not obviously or aggressively seek attention.

5. According to Coopersmith, competent, firm, accepting, and warm parenting are terms associated with the development of high _____-_____ in children.

6. The autokinetic effect is used to study the impact of group-evolved norms on _____ in children.

7. The _____ have come to share with the family the responsibility for transmitting a society's dominant cultural goals and values.

8. Children from families of higher _____ have lower dropout rates as well as lower rates of truancy, failure, and suspension.

9. The _____ legacy, which may contribute to the academic excellence of Asian-American children, emphasizes discipline and a high respect for all forms of learning.

10. Members of social groups that face a job ceiling know that they do, and this knowledge channels and shapes their children's _____ behaviors.

166

Multiple-Choice

Factual

1. a (331)
2. c (331)
3. d (332)
4. a (333-334)
5. a (334)
6. b (334)
7. b (335)
8. a (337)
9. c (338-339)
10. b (339-340)
11. b (341)
12. c (344)
13. c (346)
14. d (347)
15. c (348)

Conceptual

1. b (331)
2. a (333)
3. d (334)
4. c (335-336)
5. d (337)
6. c (337)
7. b (336-338)
8. a (338-339)
9. b (339-340)
10. d (341-342)
11. d (344)
12. a (347)
13. c (348-350)
14. d (347-351)
15. d (351)

Matching: Definitions

1. b (331)
2. l (344)
3. n (334)
4. j (334)
5. f (351)
6. p (339)
7. d (334)
8. e (346)
9. k (342)
10. c (346)
11. g (334)
12. h (346)
13. o (338)
14. m (335)
15. i (344)
16. a (341)

Completion

1. marginal (331)
2. latency (332)
3. individualistic (334)
4. popular (335)
5. self-esteem (338-339)
6. conformity (339)
7. schools (343)
8. social class (347)
9. Confucian (349)
10. academic (350)

Chapter 13

Adolescence: Developmental Acceleration and Gains

INTRODUCTION

In Chapter 13, one's biological, social, and moral foundations all go through rapid changes. The complete transition to physical and reproductive maturity occurs, with accompanying adjustments in identity, social interactions, cognitive abilities, and moral values. The following areas are specifically examined:

- Puberty. The physical developments and maturity issues that accompany the adolescent growth spurt for both boys and girls are presented. The implications of growth variations and sexual maturation are also discussed.

- Identity and Self-Concept. Adolescence is a time of egocentricity and needing to be like others. Anyone who is perceived as "different" may have difficult experiences. Recent research seems to dispute the "storm and stress" theory, although some recent studies are presented which suggest that girls emerge from adolescence with a poorer self-image than boys.

- Cognitive and Moral Conceptions. By older adolescence, most teenagers acquire the ability for logical reasoning and abstract thought, known as the period of formal operations. During adolescent transition, most teenagers also attain Kohlberg's postconventional stage of morality. Their political thinking becomes more abstract and idealistic.

- Concept of Generations. Each generation tends to fashion a somewhat unique style of thought and life--with a group consciousness and a distinct ideology.

CHAPTER OUTLINE

I. Puberty

    A. Hormonal Changes in Puberty

    B. The Adolescent Growth Spurt

    C. Maturation Among Girls

    D. Maturation Among Boys

    E. The Impact of Early or Late Maturation

    F. Self-Image

II. The Development of Identities and Self-Concepts

    A. Erikson: The Crisis of Youth

    B. Portrayal of "Storm and Stress"

    C. The Self-Concept and Self-Esteem of Adolescents

III. Cognitive and Moral Development

    A. Period of Formal Operations

    B. Adolescent Egocentricity

    C. The Adolescent as a Moral Philosopher

    D. Development of Political Thinking

    E. The Concept of Generations

LEARNING OBJECTIVES

After completing Chapter 13, you should be able to:

1. Explain the dramatic changes that occur at *puberty* by focusing on the following issues:

    a. hormonal and biological changes in adolescence (pages 356-357)

    b. *asynchrony* in growth rates (359)

2. Describe the recent theories addressing the issue that large numbers of young women in inner cities are bearing babies at younger ages: (357-358)

    a. sociobiological theory posed by Belsky, Steinberg, and Draper (1991)

    b. developmental psychology theory posed by Maccoby (1991)

    c. sociological theory

3. Explain what is meant by the *adolescent growth spurt*. (358)

4. Describe the following changes that accompany the maturity of the female reproductive system, including *menarche*: (360-361)

    a. hormonal

    b. biological

    c. cognitive

    d. social

5. Explain the following changes that accompany the maturity of the male reproductive system: (362-365)

    a. hormonal

    b. biological

    c. cognitive

    d. social

6. Discuss how early or late physical maturation can affect the self-image of boys and girls. (365)

7. Describe what Erikson calls the main task of adolescence: building and confirming a reasonable, stable identity. (366-367)

8. Cite recent research which contradicts G. Stanley Hall's theory about teenage "storm and stress." (371)

9. Explain Carol Gilligan's findings on studies of girls' self-esteem. (372-373)

10. Give examples as to why adolescence is frequently a time of storm and stress for parents. (373)

11. Describe why Piaget called adolescence the *period of formal operations*. (374-375)

12. Define what is meant by adolescent egocentrism. (378-379)

13. Discuss Kohlberg's postconventional level of morality, and tell how this affects adolescent thinking. (379-380)

14. Tell how adolescents view society and government as they make the transition through adolescence. (381)

15. Explain the "impressionable years hypothesis." (382)

KEY TERMS

---

adolescent growth spurt (page 358)

asynchrony (359)

deviant identity (369)

generation (382)

identity (366)

identity diffusion (369)

ideology (382)

menarche (361)

negative identity (369)

period of formal operations (374)

puberty (356)

puberty rites (369)

role confusion (369)

Multiple-Choice

Circle the letter of the response which *best* completes or answers each of the following statements and questions.

Factual Questions:

1. In the United States, the socially distinct period of adolescence has come about over the past 100 years because of all of the following factors *except*

    a. mandatory school attendance
    b. child labor laws
    c. theories from Freud about child abuse and neglect
    d. the fact that the nation changed from a rural to an urban society

2. The period of the life cycle when sexual and reproductive maturation becomes evident defines

    a. adolescence
    b. puberty
    c. adulthood
    d. psychological maturation

3. Which gland in the body (also known as the "master gland") secretes the hormones into the bloodstream that trigger the changes at puberty?

    a. endocrine
    b. adrenal
    c. thyroid
    d. pituitary

4. Researchers at the National Institute of Mental Health have found supportive evidence that lower levels of testosterone and higher levels of an adrenal androgen are linked to

    a. feelings of sadness and confusion
    b. good adjustment behaviors
    c. behavioral problems, including rebelliousness and fighting
    d. early menarche

5. Belsky, Steinberg, and Draper have recently advanced a controversial sociobiological theory suggesting that

    a. some young mothers are responding to an evolutionary pattern to bear children early and often
    b. boys are more likely to exhibit behavioral problems than girls
    c. youngsters growing up in affluence are likely to have more children
    d. girls reared in homes where there is a great deal of emotional stress typically enter puberty later than other girls

6. During puberty, different parts of the body grow at different rates. This is called

    a. menarche
    b. adolescent growth spurt
    c. acceleration
    d. asynchrony

7. An earlier onset of the first menstruation over the past century in Western nations appears to have resulted from

    a. genetic predispositions
    b. an improvement in overall nutrition
    c. hormone replacement therapy
    d. less strenuous lives for women

8. Early-maturing adolescent boys tend to

    a. possess feelings of adequacy in comparison with peers
    b. possess feelings of inadequacy in comparison with peers
    c. be born leaders
    d. be more aggressive and rebellious than their peers

9. Early-maturing adolescent girls tend to

    a. consider their development a severe handicap
    b. have a negatively affected prestige status
    c. be stout and develop a stocky physique
    d. all of the above

10. Which of the following statements pertaining to adolescent self-image is true?

    a. Teenagers are preoccupied with the issue of who is the most intelligent.
    b. Popularity and peer approval are major issues.
    c. Preteens are more concerned with developing intimate friendships than teenagers.
    d. Concern with weight is not an issue to an adolescent.

11. In the United States over the past fifteen years, the proportion of obese children aged 6 to 11 has

    a. decreased 25 percent
    b. increased 25 percent
    c. decreased 54 percent
    d. increased 54 percent

12. During adolescence, a state characterized by bewilderment about who one is, where one belongs, and where one is going is known as

    a. role confusion
    b. deviant identity
    c. moratorium
    d. negative identity

13. All of the following are theories about the development of the disorder anorexia nervosa *except*

    a. women desperately want to be thin and beautiful
    b. advertising and marketing agencies bombard young women with messages to "BE SLIM"
    c. the child is seeking attention from family and peers
    d. a slender body build avoids adulthood and adult responsibilities and retains one in the comfort and safety of childhood

14. Compared with American society, many cultures of the world recognize puberty through

    a. isolation experiences
    b. completion of strenuous tasks
    c. special initiation ceremonies and rites
    d. ritualistic circumcision

174

15. Recent research on the self-esteem/self-concept of adolescents has found that

   a. adolescence is a definite period of "storm and stress"
   b. adolescence is not really a period of "storm and stress" except for a small portion of this age group
   c. girls experience an increase in self-esteem and prestige during this stage
   d. boys experience a decrease in self-esteem and prestige during this stage

16. Carol Gilligan's 1991 survey of 3,000 adolescents found that

   a. girls emerge from adolescence with a poorer self-image, less confidence, and lower expectations than boys
   b. boys emerge from adolescence with a poorer self-image, less confidence, and lower expectations than girls
   c. adolescent girls tend to focus on caring only for themselves
   d. adolescent boys tend to focus on close relationships

17. Mothers and fathers report that _____ is the most difficult stage of parenting.

   a. infancy
   b. toddlerhood
   c. preadolescence
   d. adolescence

18. Jean Piaget called adolescence the period of formal operations because

   a. adolescents are able to be employed and vote in elections
   b. adolescents are formally out of the required educational system
   c. adolescents gain the ability to think in logical, abstract terms to solve complex problems
   d. adolescents can think in immediate terms

19. On the basis of information presented in the chapter on adolescent egocentricity, it would be fair to conclude that

   a. most adolescents are not concerned with what their peers think
   b. most adolescents are less self-conscious at this stage
   c. most adolescents believe other people are as admiring or critical of them as they are of themselves
   d. most adolescents are concerned with observing others rather than being observed

20. It would be fair to conclude from Kohlberg's work on moral development that

   a. by early adolescence, children are generally more liberal in their outlook
   b. adolescents are often idealistic and think they have to reform the world
   c. adolescents are often unconcerned with the adult world and its issues
   d. moral development in adolescence doesn't develop in an orderly sequence

Conceptual Questions:

1. Jennifer was an early maturer and married at 16 years of age. On the basis of Eleanor Maccoby's findings, what is a likely prediction for Jennifer's future?

   a. She will enjoy the benefits of marriage, and she will be a better parent by being young when she has children.
   b. She most likely will marry someone older and supportive of her needs.
   c. She will probably find it important to complete a high school education and go to college.
   d. She has probably made a poor choice of marital partner, and it is likely that her marriage will terminate in divorce.

175

2. Your son is disturbed because his nose seems to have grown too large and his feet and hands seem out of proportion to his legs and arms. Your best advice to him would be which of the following?

    a. Don't worry; your body dimensions aren't awkward; you're just being self-conscious and egocentric.
    b. Don't worry; early maturers like you tend to be more awkward and gangly than late maturers.
    c. Don't be alarmed; everyone in our family had big feet when they were younger, and we'll just have bigger feet when we're older.
    d. Don't worry; in kids your age the heads, hands, and feet complete their growth before the legs, arms, and trunk.

3. Your 11-year-old daughter is concerned because all her friends have started menstruating, and she hasn't. Your best advice to her would be which of the following?

    a. Don't be concerned; you've already started to grow much taller and heavier, and you'll soon be a proud young woman.
    b. There's still plenty of time; the average girl doesn't have her period until age 13.
    c. Be happy you haven't started because menstruating is a woman's lot in life and only produces discomfort.
    d. Be glad you're not menstruating because once you start you'll be capable of becoming pregnant--and we don't want that.

4. Your 16-year-old daughter is an athlete, exercises strenuously every day, and eats a very lean diet to prepare for competition. She has not started menstruating. She is alarmed by this, but your best advice to her would be which of the following?

    a. Let's get you to the doctor immediately; there's a possibility that something is wrong.
    b. Don't worry about it; it is common for this to occur with female athletes, but I'll take you for a checkup to be safe.
    c. Talk to your friends and get their advice.
    d. It doesn't matter because there'll be plenty of time for that in the future.

5. Stocky physique is to _____ as bulimia is to seriously unhealthy eating habits.

    a. anorexia nervosa
    b. athletic behavior
    c. early female maturation
    d. feelings of adequacy

6. As an adult, Jason has been an attention seeker, defies authority, has lost several jobs, and, in general, asserts unconventional adult behavior. We can infer from this description that Jason

    a. was a late maturer and hasn't made a successful transition from adolescence to adulthood
    b. was an early maturer and hasn't made a successful transition from adolescence to adulthood
    c. is more likely to have a tall, large physique
    d. is more likely to have a small physique

7. You receive a formal notice from school indicating that your son would like to join the high school fraternity and needs $50 in initiation dues. You are willing to consider this because you realize that

    a. money is no object when it comes to your child
    b. as a teenager, you know he is really concerned about being with the "popular" crowd
    c. this will probably help him establish close, intimate friendships with the most intelligent boys in school
    d. this will be an asset for him when he applies to college

8. Your adolescent daughter seems to have an obsession with being overweight. You are concerned because she has reacted to your concern about her plumpness by going on crash diets and still thinks she is overweight. Which of the following statements would be the most reasonable?

   a. You should be somewhat concerned because your daughter may be at risk for developing anorexia nervosa.
   b. Ignore the problem because her dieting is a self-conscious effort to deal with her newfound sexuality.
   c. You should advise your daughter to exercise more to burn off excessive calories.
   d. You should advise your daughter to eat fewer meals and to eat more slowly.

9. Although you want him to go to college, your teenage son expresses a desire to go into the military so that he can decide what he would really like to do with his adult life. Erikson would consider his choice an example of

   a. role confusion
   b. identity diffusion
   c. negative identity
   d. adolescent moratorium

10. John is 15, has a severe reading disability, and has been in trouble with the law for disorderly conduct and truancy from school. Erikson would suggest that John is

    a. trying to be a leader for his peers
    b. trying to fit in with a particular peer group
    c. taking on a negative identity role
    d. having difficulty adjusting to late adolescence expectations

11. An African tribe requires a young man to go out in to the forest and kill a wild boar before he is initiated into manhood in his tribe. Psychologists would suggest that this trauma

    a. will produce adult maladjustment
    b. is likely to produce an identity crisis
    c. is definitely not as healthy as the child-adolescence-adulthood transition provided in our culture in the United States
    d. is really a healthy, clearly defined transition to adulthood

12. Your daughter was a happy, healthy girl in grade school; but in high school she has become withdrawn, self-critical, and moody. On the basis of the research presented in the chapter, you can conclude that

    a. this is the typical "storm and stress" of adolescence
    b. this could be a temporary depression resulting from the fact that your family has moved frequently
    c. because of changes in puberty she is attempting to regress to childish attention-seeking behavior
    d. major reversals in self-concept occur throughout the adolescent transition

13. On the basis of the research findings presented in the chapter on adolescence, we could advise parents that

    a. this is an easier time for marriages because the kids are out of the house most of the time
    b. this is most likely the most stressful time of parenting, but teenagers still need parental support, communication, and affection
    c. youngsters are unlikely to identify with their parents and will most likely make major shifts in values
    d. the permissive style of parenting has proven to be the most successful through this stage of development

14. A young adolescent boy is presented with the following problem: "All ants that can fly are as big as zebras. This ant can fly. Is this ant as big as a zebra?" He responds by saying, "Ants can't fly; I don't know." We can conclude from Piaget's findings that this boy is

   a. in the period of concrete operations
   b. not paying attention enough to analyze the problem
   c. in the period of formal operations
   d. understanding the question but doesn't like to respond with a ridiculous answer

15. Your adolescent daughter frequently becomes moody, requests to be left alone, and then seems to bounce back from a low mood within a relatively short period of time. Should you be concerned about her mood swings?

   a. Yes, these are warning signs that she is experiencing PMS.
   b. Yes, these mood swings are early warning signs of manic depression.
   c. No, she just needs to keep her mind and hands more active by doing more household chores.
   d. No, these mood swings are a normal part of adolescent life.

16. Your teenage son stands in his closet entrance for a half hour imagining how his friends will admire him if he wears the right T-shirt to the school dance. Psychologists would explain this behavior as the

   a. personal fable
   b. stage of concrete mental operations
   c. stage of self-centered preoccupation
   d. phenomenon called adolescent egocentricity

17. Your brother and his teenage daughter are continually getting into disagreements about how his generation has made such a mess of the world. She says that her generation is going to clean up the environment, feed the homeless, and pay back the federal deficit. Your brother doesn't understand that she

   a. is learning in school that those people who grew up in the sixties and early seventies dropped out of society, and that's why it's in such a mess
   b. is naturally concerned with moral values and principles as a part of adolescent idealism
   c. is naturally going to oppose her parents because adolescents shouldn't trust anyone over 30 years of age
   d. is trying to manipulate him to get what she wants

18. An adolescent maintains that he believes in equal and fair treatment for all human beings, yet he joins a fraternity that excludes blacks from membership. This demonstrates

   a. a deficiency in his moral reasoning
   b. his adolescent egocentrism
   c. that he is in the stage of postconventional morality
   d. moral relativism

19. Joseph is increasingly concerned with questions about justice, crime, citizens' rights and obligations, and the function of government. On the basis of his inquiries, we can infer that he is most likely about

   a. 10 years old
   b. 13 years old
   c. 15 years old
   d. 18 years old

20. There may be some validity to the expression "generation gap" because

    a. parents are now in a different role than when they were adolescents
    b. parents are generally more affluent than adolescents
    c. each new generation fashions a somewhat unique style of thought and life different from that of the previous generation
    d. each generation has become less and less involved with societal issues

Matching: Definitions

Match the letter of the term in the box with its definition.

a.	asynchrony	i.	deviant identity
b.	menarche	j.	puberty rites
c.	ovulation	k.	anorexia nervosa
d.	popularity	l.	bulimia
e.	identity	m.	G. Stanley Hall
f.	role confusion	n.	Albert Bandura
g.	identity diffusion	o.	personal fable
h.	negative identity	p.	ideology

1. Elkind calls this romantic imagery of oneself the _____

2. a term used to describe the dissimilarity in the growth rates of different parts of the body _____

3. a state characterized by bewilderment about who one is, where one belongs, and where one is going _____

4. a psychologist who believes that "storm and stress" theory of adolescence is not typical _____

5. the first menstrual period, usually following the peak of the growth spurt _____

6. a lack of ability to commit oneself to an occupational or ideological position and to assume a station in life _____

7. usually does not take place until twelve to eighteen months after the first menstruation _____

8. what every adolescent strives to achieve with peers during this stage of development _____

9. a disorder in which one willfully suppresses appetite, resulting in self-starvation _____

10. the main task of adolescence, according to Erikson, is to develop this _____

11. published *Adolescence* in 1904, depicting this period as a time of "storm and stress" _____

12. what Erikson says is a debased self-image and social role _____

13. a set of shared definitions that offer interpretations and solutions to what is felt to be an unsatisfactory social condition _____

14. a disorder also called the binge-purge syndrome _____

15. initiation ceremonies that socially symbolize the transition from childhood to adulthood _____

16. a lifestyle that is at odds with, or not supported by, the values and expectations of society _____

Completion

Supply the term or terms needed to complete each of the following statements.

1. _____ is the period of the life cycle when sexual and reproductive maturation becomes evident.

2. _____ factors also have consequences for social relations during adolescence.

3. The links between hormone levels and behavior tend to be considerably stronger and more consistent for _____ than for _____.

4. _____ theory contends that pubertal timing is an outcome of social experience and evolutionary factors.

5. Rapid increase in height and weight during the adolescent years is called the _____ _____ _____.

6. The earlier onset of menarche appears to be caused largely by _____ improvements.

7. Any variation from the peer group in growth and development tends to affect the adolescent's _____ _____.

8. The proportion of children aged 6 to 11 who are obese has increased _____ percent over the past fifteen years.

9. _____ work has focused attention on the struggle of adolescents to develop and clarify their identity.

10. Piaget called adolescence the period of _____ _____, the final and highest stage in the development of intelligence.

Multiple-Choice

Factual

1.	c	(356)
2.	b	(356)
3.	d	(356)
4.	c	(357)
5.	a	(357)
6.	d	(359)
7.	b	(361)
8.	a	(364)
9.	d	(364-365)
10.	b	(365)
11.	d	(366)
12.	a	(369)
13.	c	(370)
14.	c	(371)
15.	b	(371-372)
16.	a	(372)
17.	d	(373)
18.	c	(374)
19.	c	(378-379)
20.	b	(379)

Conceptual

1.	d	(358)
2.	d	(359)
3.	b	(361)
4.	b	(361)
5.	c	(363-364)
6.	a	(364)
7.	b	(365)
8.	a	(366-367)
9.	d	(368)
10.	c	(369)
11.	d	(369-371)
12.	b	(372)
13.	b	(373)
14.	a	(374)
15.	d	(375)
16.	a	(378)
17.	b	(379)
18.	d	(380)
19.	d	(381)
20.	c	(382)

Matching: Definitions

1.	o	(378)
2.	a	(359)
3.	f	(369)
4.	n	(371)
5.	b	(361)
6.	g	(369)
7.	c	(361)
8.	d	(365)

9.	k	(370)
10.	e	(366)
11.	m	(369)
12.	h	(369)
13.	p	(382)
14.	l	(370)
15.	j	(369)
16.	i	(369)

Completion

1.	puberty (356)	
2.	biological/hormonal (357)	
3.	boys; girls (357)	
4.	sociobiological (357)	
5.	adolescent growth spurt (358)	

6.	nutritional (361)
7.	self-image (365)
8.	54 (366)
9.	Erikson's (367)
10.	formal operations (374)

Chapter 14

Adolescence: Threshold of Adulthood

INTRODUCTION

Chapter 14 discusses the stage of adolescence in the context of society's need to facilitate young people's quest for autonomy and a positive self-identity. Several significant issues are explored, including the following:

- Peer Groups. The unique developmental role of the adolescent peer group is discussed, as well as the importance of adolescents' relationships with their families.

- Teenage Sexuality. The difficult adjustment by young people in regard to their experiences with dating, love, and sexuality is examined. The typical patterns and functions of dating are presented; the indefinable notion of love, as well as a possible physiological basis for this romantic attraction, is discussed. The chapter reviews the dynamic state of adolescent sexual expression, recent changes in sexual behavior, and (within this context) problems associated with teenage pregnancy.

- Vocational Options. The importance of adolescent decisions regarding vocational options is presented, as well as the issue of working teens.

- Other Significant Issues. Other important issues relevant to teenagers are examined, such as drug abuse, teenage suicide, juvenile delinquency, and high school attrition.

I. Peers and Family

    A.   The Adolescent Peer Group

    B.   Adolescents and Their Families

II. Dating, Love, and Sexuality

    A.   Dating

    B.   Love

    C.   Sexual Attitudes and Behavior

    D.   New Codes of Sexual Morality

    E.   Teenage Pregnancy

III. Vocational Choice

IV. Some Special Issues

    A.   Drug Abuse

    B.   Teenage Suicide

    C.   Juvenile Delinquency

    D.   High School Dropouts

LEARNING OBJECTIVES

After completing Chapter 14, you should be able to:

1.   Explain the relationship between the following concepts: (page 390)

    a.   *generation gap*

    b.   *youth culture*

    c.   *consciousness of oneness*

2.   Describe the makeup of the typical youth culture as well as the qualities needed to maintain a positive status within this group. (390-391)

3. Discuss the developmental phases and functions of adolescent peer groups. (391-392)

4. Explain the influence of adolescents' families on their behavior. (392-393)

5. Distinguish between the different kinds of experiences provided to adolescents by peers versus families. (393)

6. Discuss the relationship between adolescents and their families in regard to the shifting power equation within the family. (393-394)

7. Describe adolescent girls' relationships with their mothers and girlfriends. (Box, 394-395)

8. Cite the influence that each of the following factors exerts on adolescent behavior: (394-397)

   a. sexual attraction

   b. dating

   c. adolescent views of love

9. Summarize the research on the following sexual behaviors: (398-400)

   a. masturbation

   b. nocturnal orgasm

   c. heterosexual petting and intercourse

   d. homosexual activity

10. Cite the research findings regarding the impact of AIDS on adolescent perceptions and behaviors. (400-401)

11. Critically appraise the newly emerging code of sexual morality in adolescence, taking into consideration the data on adolescent contraceptive use and pregnancy. (401-403)

12. List the factors that influence adolescent vocational choice. (403-404)

KEY TERMS
---

consciousness of oneness (page 390)

drug abuse (406)

generation gap (390)

youth culture (390)

SELF-TESTS - CHAPTER 14

Multiple-Choice

Circle the letter of the response which *best* completes or answers each of the following statements and questions.

Factual Questions:

1.   When social scientists talk about a generation gap, they are referring to

   a.   mutual antagonism and separation between young people and adults
   b.   the youth-oriented and age-fearing culture we live in
   c.   the age-segregated nature of most family activities
   d.   a youth-culture idea that the past is irrelevant and the future is uncertain

2.   The youth culture tends to emphasize _____ as a means of achieving status.

   a.   drug abuse
   b.   juvenile delinquency
   c.   confidence in one's sexual identity
   d.   a consciousness of oneness

3.   Social scientists have concluded that the

   a.   youth culture is a monolithic group for all adolescents
   b.   youth culture is composed of several crowds that often display mutual antagonism
   c.   peer group has a greater influence on the adolescent than the family
   d.   differences among young people are greater than their similarities

4.   Adolescents are most influenced by their parents when the issues pertain to

   a.   career plans
   b.   personal adornment
   c.   autonomy
   d.   communication

186

5. Between childhood and adolescence, sons

    a. acquire power at the expense of their mothers
    b. become less assertive in dealing with their parents
    c. replace their father in the influence hierarchy of the family
    d. move into a lower overall position in the family influence hierarchy

6. Which of the following is *not* a function served by adolescent dating?

    a. fostering heterosexual relations
    b. replacing the influence of the nuclear family
    c. exploring mutual compatibility
    d. companionship and problem sharing

7. According to Zick Rubin, the three components of romantic love are

    a. caring, knowing, and responsibility
    b. respect, knowledge, and caring
    c. infatuation, knowledge, and eroticism
    d. caring, attachment, and intimacy

8. Research on adolescent sexual expression indicates that

    a. boys first experience nocturnal orgasms in their late teens
    b. group masturbation is uncommon among preteen boys
    c. teenage masturbation is often accompanied by erotic fantasies
    d. sex play with other children begins in adolescence

9. Evidence suggests that, with regard to sexual behavior in the United States, this century witnessed a pattern of very rapid change

    a. around the time of World War I and during the Vietnam War years
    b. during the years of the Great Depression (in the 1930s)
    c. during each war period: World War I, World War II, the Korean War, and the Vietnam War
    d. during periods of economic depression

10. The new sexual morality is an attempt to

    a. condone promiscuous thrill seeking
    b. view sexual intercourse as a victimless act
    c. hold men to different standards than women
    d. realize traditional values of honesty and openness

11. The decision of a young couple to engage in sexual relations is increasingly being judged as morally acceptable if

    a. they are out on their first date
    b. they are emotionally involved with each other
    c. affection is absent
    d. they are "looking for a good time"

12. High school sophomores who work more than fifteen hours a week are at a great risk for

    a. committing suicide
    b. dropping out
    c. becoming delinquent
    d. having unstable marriages

13. Among adolescents in the United States, the most frequently abused drug is

    a. alcohol
    b. marijuana
    c. cocaine
    d. heroin

14. Data on suicide reveal that more

    a. women attempt and complete suicide than men
    b. men attempt and complete suicide than women
    c. women attempt suicide than men, but more men actually succeed in their attempts
    d. men attempt suicide than women, but more women actually succeed in their attempts

15. Sociologist Delbert Elliott studied school dropouts. He found that they

    a. commit fewer offenses after leaving school than in a comparable period before they
       dropped out
    b. commit more offenses after leaving school than in a comparable period before they
       dropped out
    c. have better self-images than students who stay in school
    d. have lower delinquency rates than those students who stay in school

Conceptual Questions:

1. Jill is the captain of the girls' swim team. She is attractive, socially poised, and popular with
   the "in crowd" in her school. Which of the following conclusions would be *least* appropriate?

    a. She would be a youth culture celebrity.
    b. Her self-esteem should be higher than that of her peers.
    c. Her characteristics would not have served her well if she grew up in her mother's
       generation.
    d. Her characteristics may make others wish for a consciousness of oneness with her.

2. Most modern high schools have groups of students referred to as "jocks," "nerds,"
   "burnouts," and "brains." This provides evidence of

    a. a youth culture
    b. a monolithic peer group
    c. heterogeneous peer groups
    d. heterosexual cliques

3. Your adolescent son has started to wear an earring in one ear like some of the actors on the
   television show *90210*. You should therefore conclude that his behavior indicates a(n)

    a. deviant identity
    b. effort to renounce your values
    c. conformist trend he'll overcome in high school
    d. normal form of peer-group identification

4. When you ask your 14-year-old son to attend a weekend family reunion, he refuses, saying, "They're just a bunch of old fogies; I'd rather stay here and hang out with my friends." The most reasonable conclusion you can reach is that

    a. the youth culture is more important to your son than his family
    b. if your son hung around with another group of friends, he would be more likely to respect your wishes
    c. your son still loves and respects you, but he'd rather do things because he wants to than because he's expected to
    d. by tenth grade, your son's peer group will have less influence on him, and your family generation gap will be less

5. You are a female high school guidance counselor. One of your female students approaches you and verbalizes a concern about not being normal because she is 16 years old, is still a virgin, and has not yet begun dating. Your best advice to her would be which of the following?

    a. Virgins like yourself are psychologically healthier than other girls because you believe that romantic love should precede rather than follow sex.
    b. Being a virgin isn't abnormal; maybe you're looking for a guy similar to yourself.
    c. Dating is just game playing; people like yourself concentrate on your studies and are better off in the long run.
    d. Consider masturbation while imagining erotic scenes.

6. Your 11-year-old son asks you, "If a guy masturbates, does that make him a homosexual?" Your *best* response to him would be that masturbation is

    a. normal, but doing it with friends is abnormal
    b. OK as long as you don't have impure thoughts while doing it
    c. a healthy sexual outlet as long as you don't feel guilty about it
    d. not harmful, but homosexual play can lead to being a homosexual

7. You are developing a sex education curriculum for junior high students. If you were to provide these students with an accurate description of current norms for heterosexual activity, you could say which of the following?

    a. If it feels good, do it.
    b. What's sauce for the gander should be sauce for the goose.
    c. Just think twice before you say it is all right.
    d. Make new friends but keep the old ones.

8. On the basis of recent research on adolescent sexual behavior and attitudes, your adolescent daughter would probably feel that

    a. it is all right for adolescent boys but not girls to have sexual intercourse
    b. teens should feel free to engage in sexual intercourse if they are in love and plan to marry
    c. the decision to have intercourse is a private matter that should not be discussed publicly
    d. adolescents should have complete sexual freedom as long as they are well adjusted

9. A teenage girl comes into a federally funded birth control clinic for her first visit. Which of the following assumptions would be the *most* valid?

    a. She has probably never used any form of contraception.
    b. She started using contraception months after she became sexually active.
    c. She thinks that if she does not want to become pregnant, she won't.
    d. She is probably using withdrawal as a contraceptive technique.

10. Part-time employment is to social responsibility as racism is to _____.

    a. unemployment
    b. gatekeepers
    c. employability
    d. personal responsibility

11. You caught your son smoking pot in his bedroom. You can legitimately conclude that

    a. your use of sleeping pills did not influence his decision to smoke pot
    b. he perceives a consciousness of oneness with you because you drink cocktails when you get home from work
    c. his pot smoking makes him feel "cool" in competitive situations and increases his status
    d. regular use of pot may adversely affect his future school performance

12. Your daughter informs you that she sometimes drinks alcohol and smokes pot at parties. According to recent research on casual drug experimentation, what is the most valid conclusion you can draw about your daughter's disclosure?

    a. It will lead to addiction.
    b. It is associated with poor social adjustment.
    c. It will lead to pathological functioning.
    d. It is frequently associated with good social and personal adjustment.

13. A teenage relative of yours has just made a suicide attempt. Which of the following conclusions would be *least* valid?

    a. The relative is more likely to be male than female.
    b. Feelings of helplessness and boredom preceded the attempt.
    c. The relative was ambivalent about his or her willingness to die.
    d. The relative had an increased number of psychosomatic complaints.

14. Which of the following adolescents would you be *most* concerned about? A boy who

    a. normally has a B+ overall average but is now failing three of his classes
    b. considers school a waste of time and likes hanging around with jobless dropouts
    c. is not interested in grades but cares only about extracurricular activities
    d. is emotionally immature and enjoys playing the class clown

15. Jimmy is in the ninth grade at Mulberry High School. He has sixth-grade equivalent scores in reading and math, has a long truancy record, and resents his teachers hassling him. We are likely to infer that Jimmy will

    a. go to a vocational school and gain employment skills
    b. impulsively attempt suicide to escape his helpless situation
    c. stay out of mischief if he is retained in school
    d. drop out of high school

Matching: Definitions

Match the letter of the term in the box with its definition.

a.	petting	e.	love
b.	masturbation	f.	consciousness of oneness
c.	youth culture	g.	drug abuse
d.	suicide	h.	generation gap

1. associated with the expression of attachment, caring, and intimacy _____

2. the existence of mutual antagonism, misunderstanding, and separation between young people and adults _____

3. the second or third leading cause of death among adolescents _____

4. a form of erotic self-stimulation that may occur through fondling of the penis or the manual stimulation of the clitoris _____

5. the excessive or compulsive use of chemical agents to an extent that it interferes with people's health, social or vocational functioning, or the functioning of the rest of society _____

6. a sympathetic identification in which group members come to feel that their inner experiences and emotional reactions are similar _____

7. more or less standardized ways of thinking, feeling, and acting that are characteristic of a large body of young people _____

8. any form of erotic caressing that may or may not lead to orgasm _____

Completion

Supply the term or terms needed to complete each of the following statements.

1. Two qualities that are essential for obtaining high status in the adolescent youth _____ are projecting an air of confidence in one's sexual identity and the ability to deliver a smooth performance.

2. Every high school typically has several _____ that are often mutually exclusive and antagonistic to one another.

3. Between childhood and adolescence, _____ tend to acquire power at the expense of their _____.

4. The term _____ has itself become, in many ways, too stiff and formal to describe the "just hanging out" and "getting together" that takes place among contemporary youth.

5. Half of America's young people have had sexual intercourse by the time they are _____ years old.

6. Many young people have come to judge the acceptability of sexual behavior in terms of a couple's _____ involvement.

7. Although it is commonly assumed that _____ provides youth with on-the-job training of a _____ useful nature, it actually affords very little.

8. A factor in the use of illegal drugs by young people is that they view their parents as users of _____ agents such as tranquilizers, barbiturates, and stimulants.

9. A dramatic decline in the quality of an adolescent's schoolwork is often considered to be a _____ warning sign.

10. High school _____ have often failed at least one grade and are behind in reading and math performance.

ANSWERS FOR SELF-TESTS - CHAPTER 14

Multiple-Choice

Factual

1.	a	(390)
2.	c	(391)
3.	b	(391-392)
4.	a	(392)
5.	a	(393-394)
6.	b	(396)
7.	d	(397)
8.	c	(398)
9.	a	(398)
10.	d	(401)
11.	b	(401)
12.	b	(405)
13.	a	(406)
14.	c	(408)
15.	a	(411)

Conceptual

1.	c	(390-391)
2.	c	(391-392)
3.	d	(392)
4.	c	(393)
5.	b	(395-396)
6.	c	(398)
7.	b	(399-400)
8.	b	(401)
9.	b	(402)
10.	c	(403-404)
11.	d	(406)
12.	d	(406-407)
13.	a	(408)
14.	a	(409)
15.	d	(411)

Matching: Definitions

1.	e	(397)
2.	h	(390)
3.	d	(408)
4.	b	(398)

5.	g	(406)
6.	f	(390)
7.	c	(390)
8.	a	(398)

Completion

1.	culture (391)
2.	crowds (392)
3.	sons, mothers (393-394)
4.	dating (396)
5.	17 (399-400)

6.	emotional (401)
7.	work; technically (404)
8.	psychoactive (408)
9.	suicide (409)
10.	dropouts (411)

Chapter 15

Early Adulthood: Perspectives and Development

INTRODUCTION

Chapter 15 focuses on the dynamic life stage of young adulthood. Developmental perspectives of adulthood are introduced--illustrating that this period of life is a process of "becoming," as opposed to a static state of being. Several relevant topics are emphasized, including:

- Different Perceptions of Adulthood. Within this context, demographic features of this age group are revealed, and conceptions of the various age periods are discussed. Biological and social changes that accompany adulthood are reviewed. The functions of age-grade systems are presented, along with an explanation of how these serve to prepare individuals for the various roles associated with each distinct life stage. Finally, life events--those critical turning points at which people change direction in the course of their lives--are examined.

- Typical Periods and Transitions. Erikson's psychosocial stage theory is discussed within this setting. Levinson's perspective, which focuses on particular developmental tasks confronting men throughout adult development, is also presented. Research on the distinct phases which accompany adult female development is examined in depth. Criticisms of the stage approach to adult development are outlined.

- Physical Changes and Health. The implications of the typical changes in adult physical performance are illustrated, and changes in mental health are also highlighted.

- Changes in Cognitive Development. A brief discussion occurs on the possibility of a "post-formal stage" of cognitive development. Sternberg's "information-processing" approach explains how adults conceptualize their complex world. Finally, adult moral reasoning capacity is discussed.

CHAPTER OUTLINE

I. Developmental Perspectives

   A. Demographic Aspects of Adulthood

   B. Conceptions of Age Periods

   C. Age Norms and the Social Clock

   D. Age-Grade Systems

   E. Life Events

II. The Search for Periods in Adult Development

   A. Erikson: Psychosocial Stages

   B. Levinson: Phases in Adult Male Development

   C. Phases in Adult Female Development

   D. The Stage Approach Controversy

III. Physical Changes and Health

   A. Physical Performance

   B. Physical Health

   C. Mental Health

IV. Cognitive Development

   A. Post-Formal Operations

   B. Thought and Information Processing

   C. Moral Reasoning

LEARNING OBJECTIVES

After completing Chapter 15, you should be able to:

1. List the particular demographic features of adulthood. (pages 416-417)

2. Summarize the various conceptions that people have of the different age periods of life. (417-419)

3. Define the following and identify the relationship among each: (419-420)

   a. *aging*

   b. *biological aging*

   c. *social aging*

   d. *social norms*

   e. *age norms*

   f. *social clock*

4. Describe the functions that age-grade systems provide for adults across different cultures. (421-422)

5. Discuss the role that *life events* play in helping people locate themselves across the life span. (422-424)

6. Explain what is known about rape as a stressful life event. (Box, 423)

7. Summarize the major theme of Erikson's approach to adult development. (424-425)

8. List the various levels of Levinson's phases of adult male development. (426-427)

9. Describe the new social definitions of adult women. (428)

10. Summarize the reasons why students' temporarily drop out of college. (Box, 428)

11. Tell the significance of the following phases of adult female development: (428-433)

   a. family and work

   b. reentering the paid labor force

   c. stocktaking

   d. unique female experiences

   e. depression

12. Explain what opportunities exist for professional athletes once their careers are over. (Box, 430)

13. Discuss stage approaches to adult development. (433-434)

14. List the changes that occur in each of the following topics in adulthood: (435-440)

    a. physical performance

    b. physical health

    c. mental health

15. Describe what is known about adult leisure activities. (Box, 438-439)

16. Give examples of post-formal mental operations, adult *information processing*, and adult moral reasoning. (440-445)

17. Discuss how adults deal with stressful life events. (442-443)

KEY TERMS

age norm (page 419)

aging (419)

biological aging (419)

information processing (441)

life event (422)

social aging (419)

social clock (420)

social norm (419)

thinking (441)

SELF-TESTS - CHAPTER 15

Multiple-Choice

Circle the letter of the response which *best* completes or answers each of the following statements and questions.

Factual Questions:

1. The notion that adulthood is a state of being has increasingly come to be replaced by a view of adulthood as a process of

   a. surviving
   b. becoming
   c. stabilization
   d. reductionism

2. Because of the size of the baby-boom generation

   a. the nation's labor force has rapidly expanded
   b. competition for professional jobs has decreased
   c. the number of teenagers has increased
   d. these individuals now make up one-half of the American population

3. Older adults are viewed _____ favorably and as _____ desirable to be around than younger adults.

   a. more, more
   b. more, less
   c. less, less
   d. less, more

4. Changes in the structure and functioning of the human organism through time best defines

   a. a social clock
   b. biological aging
   c. the BOOM phenomenon
   d. cultural adaptation

5. People pass through a socially regulated cycle in their assumption and relinquishment of roles through time. This is called

   a. biological aging
   b. social aging
   c. the BOOM phenomenon
   d. resettlement

6. Although the members of a society tend to share similar expectations about the life cycle, some variations do occur. The social clock tends to be paced more rapidly in the

   a. upper-upper class
   b. lower-upper class
   c. middle class
   d. lower class

7. The arrangement of people in social layers that are based on periods in the life cycle best defines

   a. an age-grade system
   b. the empty nest
   c. biological aging
   d. the BOOM phenomenon

8. Fires, storms, and tidal waves are _____ that derive from events in the physical world.

   a. life stages
   b. life events
   c. developmental stressors
   d. constructed realities

9. According to Daniel Levinson and his associates, a man commonly has a sense of disparity between "what I've reached at this point" and "what it is I really want" that leads him to an interval of soul searching during which of the following life periods?

   a. restabilization
   b. getting into the adult world
   c. the mid-life transition
   d. leaving the family

10. Women typically reappraise the relative importance of family and career during the

    a. stocktaking period
    b. boom decade
    c. restabilization period
    d. age-30 transition

11. Bernice Neugarten believes that a stage approach oversimplifies adult life since

    a. many inner changes occur slowly across the life span and not in a steplike fashion
    b. the timing of life events is becoming more regular
    c. the psychological themes reported by adults of each age group are different ones
    d. adult stages form a unique sequence or order

12. Research shows that, compared with nonsmokers, adult smokers are more likely to be

    a. introverted
    b. inhibited
    c. compliant
    d. risk takers

13. The most common mental ailment of adult Americans today is

    a. depression
    b. schizophrenia
    c. alcohol abuse
    d. anxiety disorder

14. An adult who can accept mutually incompatible systems of knowledge is displaying

    a. post-formal operations
    b. formal operations
    c. information processing
    d. insightful learning

15. Stress-resistant individuals display an openness to change, feelings of involvement in what they are doing, and

    a. lowered psychological arousal
    b. heightened sensitivity to other people
    c. more life events experience
    d. a sense of control over events

Conceptual Questions:

1. Which of the following descriptions best represents the prevailing attitude of young adults who were born in the late 1960s and early 1970s?

    a. idealistic
    b. family-oriented
    c. materialistic
    d. altruistic

2. You always joke around with your 25-year-old, unmarried sister and tell her she's two and a half decades old. She is likely to

   a. take you seriously and feel older than she actually is
   b. ignore your joking and be happy with her age
   c. speed up her social clock by getting married
   d. employ a double standard of aging when viewing her own age

3. A year after your father died, your elderly mother started living with an elderly gentleman in her senior citizens' apartment complex. Now, no one in your mother's building talks to her. This example demonstrates the effect of

   a. violating the social clock
   b. violating age norms
   c. ageism and sexism directed at women
   d. conflict between age grades

4. Shirley grew up in a poor coal-mining town in Appalachia. She got married at the age of 15 and had three children by the time she was 20. We can infer that

   a. she will feel younger because she'll still be fairly young when her children are all grown
   b. Shirley's social clock has been slowed down by her early introduction to adulthood
   c. she will feel older than she is because her social clock has been accelerated
   d. Shirley would experience role confusion because there was no clear transition between her childhood and adulthood

5. In the country of Castlana, young men are prevented from owning private property until they reach the age of 45. However, under the influence of Western political ideas, young men in Castlana are attempting to change their _____ political system into a more democratic one.

   a. age-grade
   b. socially ageist
   c. socially clocked
   d. age stereotyped

6. Your grandfather always talks about the war years and about how he went to college on the G.I. Bill and then married your grandmother. We can infer that

   a. his reminiscing accomplishes the task of generativity
   b. your grandfather is living in the past because his memories are more pleasant than the "here and now"
   c. your grandfather has confused the biological aging process with the social aging process
   d. these life events help your grandfather's sense of accomplishment

7. Which of the following examples would be consistent with a stage theory of adult development?

   a. A person's social identity is continually affected throughout the life cycle by marriage, divorce, unemployment, and other events.
   b. As adults develop, they gradually become less active and less adaptive.
   c. Healthy adult development can occur no matter when a person marries, has children, leaves the family, and so on.
   d. Young adults tend to be impulsive and adaptive, whereas older adults are deliberate and less flexible.

8. Your 19-year-old son has not decided whether he wants to join the army or go to college. Research concerning stages of adult male development predicts that

   a. his major concern is in gaining financial self-sufficiency and gaining autonomy from the family
   b. he is most concerned with starting a career and creating a dream about his future
   c. he is obsessed with becoming his own man and competing for promotions
   d. he is experiencing inner doubts about who he is and where he is going

9. Which of the following is *most* accurate regarding new social definitions of adult females?

   a. Since 1890, proportionately more white women have been employed outside their homes than nonwhite women.
   b. Women between the ages of 50 and 65 were more likely to be associated with chronological age than with phases of the family cycle.
   c. Women are still largely perceived in regards to their reproductive role.
   d. Employment outside the home is playing an increasingly important role in the self-esteem and identity of women.

10. Your wife is 30 years old and has been a childless traditional homemaker for the last seven years. Which of the following predictions would be the *most* valid?

    a. She is likely to perceive herself as being older than she actually is.
    b. She is likely to conform to traditional age and social norms.
    c. The stocktaking she is performing may lead to symptoms of depression.
    d. She is likely to take on a mentor who will encourage her to return to the labor force.

11. You are arm wrestling with your 20-year-old son. Which of the following conclusions would be the *most* valid?

    a. He will win because his arm strength is at its maximum.
    b. You will win because you are a construction worker, and he is a sedentary student.
    c. He will win because quickness and agility can overcome your greater strength.
    d. Although his physical strength should be at its peak, we still cannot predict a clear winner because of the minimal changes in physical performance which accompany adult development.

12. A biologist accepts the principle of determinism, assuming that events have causes which can be established experimentally. Yet the same biologist believes in God, a belief that cannot be proven experimentally. This biologist is demonstrating one of the aspects of

    a. post-formal mental operations
    b. mental rigidity
    c. dialectical reasoning
    d. information encoding

13. Which of the following statements is *not* in accordance with the findings regarding adult mental health?

    a. Mental health involves the ability of people to function effectively in their social roles.
    b. It entails a subjective sense of well-being.
    c. It is enhanced when people remain static throughout life's challenges.
    d. It requires that people continually change and adapt to life's fortunes.

14. In computer programming courses, the student is often asked to write a program solution in pseudocode (e.g., an English-language description of what the computer must do to solve the problem) before writing the actual programming language. The student following these instructions would be using

   a. concrete mental operations
   b. insightful learning
   c. encoding of information
   d. post-conventional reasoning

15. A group of students are taking their final exam. Which of the following students is likely to perform best in this situation?

   a. a student who has elevated levels of the hormone ACTH throughout the exam
   b. a student who says to himself, "Boy, I could flunk this test, but why think of that? I'll do the best I can"
   c. a student who says to herself, "I don't know what I'll do if I don't pass this exam"
   d. a student who has recently dealt with other stressful events

Matching: Definitions

Match the letter of the term in the box with its definition.

a.	aging	i.	social aging
b.	restabilization	j.	thinking
c.	role overload	k.	social clock
d.	biological aging	l.	life event
e.	social norm	m.	age-30 transition period
f.	information processing	n.	alarm reaction
g.	age norm	o.	role conflict
h.	social support	p.	age-grade system

1. changes in an individual's assumption and relinquishment of roles through time _____

2. a stage in Levinson's theory where the mid-life transition comes to an end and a new life structure begins to take shape _____

3. a turning point at which individuals change some direction in the course of their lives _____

4. changes in the structure and functioning of the human organism through time _____

5. occurs when women experience pressures within one role that are incompatible with the pressures that arise within another role _____

6. a process by which we change and reorganize information to create new information _____

7. the arrangement of people in social layers that are based on periods in the life cycle _____

8. a social expectation that specifies what constitutes appropriate and inappropriate behavior for people at various periods in the life span _____

9. occurs when women have too many role demands and too little time to fulfill them _____

10. a time during which women tend to reverse the priorities they established when they were in their twenties _____

11. a set of internalized concepts that regulate our progression through the age-related milestones of the adult years _____

12. the first response to stress, during which the nervous system is activated, digestion slows, and both heartbeat and blood pressure increase _____

13. biological and social change across the life span _____

14. the step-by-step mental operations that we use in tackling intellectual tasks _____

15. consists of the exchange of resource among people based on their interpersonal ties _____

16. a standard of behavior that members of a group share and to which they are expected to conform _____

Completion

Supply the term or terms needed to complete each of the following statements.

1. Developmental psychologists currently view adulthood as a process of _____, a stage of life that involves negotiating ups and downs and changing direction to surmount obstacles.

2. Age norms are most obvious when they are embodied in _____ rules or explicit policies about a role.

3. The lower the _____ _____, the more rapid the pacing of the social clock tends to be.

4. On the surface, societies with _____-_____ systems seem to provide an orderly method for role allocation and reallocation.

5. According to Levinson, a man tends to give up his _____ during the middle to late thirties.

6. Until relatively recently, a woman's life was seen primarily in terms of her _____ role.

7. The disproportionate rate of _____ among women seems to be related to cultural traditions that diminish feminine worth and induce passive, dependent personality patterns and negative thinking.

8. Alcohol is not a _____ but a tranquilizer or relaxant.

9. Sternberg found that the best problem solvers spend more time on _____ than the poor problem solvers.

10. During the _____ stage of responding to stress, the body mobilizes its resources to overcome the stress.

Multiple-Choice

Factual

1.	b	(416)
2.	a	(416-417)
3.	c	(418)
4.	b	(419)
5.	b	(419)
6.	d	(420)
7.	a	(421-422)
8.	b	(422)
9.	c	(427)
10.	d	(429)
11.	a	(433)
12.	d	(436)
13.	c	(437)
14.	a	(440)
15.	d	(442)

Conceptual

1.	c	(417)
2.	a	(418)
3.	b	(419)
4.	c	(420)
5.	a	(421-422)
6.	d	(422-424)
7.	d	(424)
8.	a	(426)
9.	d	(428)
10.	c	(431-433)
11.	d	(435)
12.	a	(440)
13.	d	(440)
14.	c	(441)
15.	b	(442-443)

Matching: Definitions

1.	i	(419)
2.	b	(427)
3.	l	(422)
4.	d	(419)
5.	o	(430)
6.	j	(441)
7.	p	(421)
8.	g	(419)

9.	c	(430)
10.	m	(429)
11.	k	(420)
12.	n	(442)
13.	a	(419)
14.	f	(441)
15.	h	(444)
16.	e	(419)

Completion

1.	becoming (416)
2.	formal (419)
3.	social class (420)
4.	age-grade (421)
5.	mentors (427)

6.	reproductive (428)
7.	depression (433)
8.	stimulant (435)
9.	encoding (441)
10.	resistance (442)

Chapter 16

Early Adulthood: Love and Work

INTRODUCTION

Chapter 16 continues the exploration of early adulthood with the primary focus on love and work--the central theme of adult life. The chapter is subdivided into three main categories:

- Options in Lifestyles. A variety of lifestyle patterns are examined: being single, living together, young adults living at home, and marriage. Homosexuality and bisexuality are discussed as well, and several theories are presented regarding the sources and causes of these behaviors.

- The Family Life Cycle and Parenthood. Sociologist Reuben Hill's nine stages in the family life cycle are outlined, and the research is critically evaluated. Pregnancy and the beginning of parenthood are discussed, with emphasis on the changes parents and families face during these transitional periods. Various current lifestyle patterns are examined as well, including those of employed mothers, single-parent mothers, and single-parent fathers.

- Work. The significance of work for adults is discussed, including the social as well as the economic functions of work. The socioeconomic life cycle technique for studying the course of an individual's occupational status attainment is presented according to the study of both sociologists and psychologists.

I. Options in Lifestyles

    A. Singles

    B. Living Together

    C. Young Adults Living at Home

    D. Homosexual and Lesbian Couples

    E. Marriage

II. The Family Life Cycle and Parenthood

    A. Stages in the Family Life Cycle

    B. Pregnancy

    C. Transition to Parenthood

    D. Employed Mothers

    E. Single-Parent Mothers

    F. Single-Parent Fathers

III. Work

    A. The Significance of Work

    B. Status Attainment: The Socioeconomic Life Cycle

LEARNING OBJECTIVES

After completing Chapter 16, you should be able to:

1. Define the following terms, and explain how each influences the choice of relationships and *lifestyles* in adulthood: (page 450)

    a. *expressive tie*

    b. *primary relationship*

    c. *instrumental tie*

    d. *secondary relationship*

2. Evaluate the research regarding the following patterns of living, and discuss the advantages and disadvantages of each: (450-455, 458)

    a. being single

    b. living together

    c. living at home with parents

    d. *homosexual*, lesbian, and *bisexual relationships*

3. Summarize the data and theories regarding the origins of *homosexuality*. (Box, 456-457)

4. Define *marriage*, and describe the ways in which societies differ in how they structure marriage relationships. (458-459)

5. Describe the stages in the *family life cycle*, including pregnancy, transition to parenthood, and difficulties accompanying each stage. (460-465)

6. Develop an opinion on the issue of whether or not the American family is disintegrating. (Box, 462-463)

7. Examine the variety of explanations for the postpartum blues. (Box, 466)

8. Discuss the literature dealing with the effects of maternal employment, single-parent mothers, and single-parent fathers. (466-472)

9. Summarize the literature regarding the three theories advanced by sociologists to explain poverty. (Box, 468-469)

10. Explain the significance of work for adults. (473-474)

11. List the influential factors in the stages of the *socioeconomic life cycle*, and discuss the use of the *path-analysis* procedure for determining status attainment. (474-475)

KEY TERMS

---

bisexuality (page 454)

expressive tie (450)

family life cycle (460)

heterosexuality (454)

homosexuality (454)

instrumental tie (450)

lifestyle (450)

marriage (458)

path analysis (475)

primary relationship (450)

secondary relationship (450)

social relationship (450)

socioeconomic life cycle (474)

SELF-TESTS - CHAPTER 16

Multiple-Choice

Circle the letter of the response which *best* completes or answers each of the following statements and questions.

Factual Questions:

1. A social link formed when we cooperate with another person to achieve a limited goal best describes a(n)

    a. expressive tie
    b. primary relationship
    c. life cycle
    d. instrumental tie

2. _____ are to expressive ties as _____ are to instrumental ties.

    a. primary relationships; secondary relationships
    b. social relationships; lifestyles
    c. secondary relationships; primary relationships
    d. homosexual relationships; heterosexual relationships

3. Which of the following statements regarding singles is the *least* accurate?

   a. Single status among both men and women under 35 years of age has decreased in recent years.
   b. A growing proportion of Americans will never marry.
   c. The population remaining single today is smaller than it was at the turn of the century.
   d. Single men and women were ridiculed and treated with disapproval during colonial days.

4. One of the major advantages of being single is

   a. achieving maximum happiness
   b. greater freedom and independence
   c. the lessening of social stigma
   d. the impersonal nature of relationships

5. Living together as an alternative to marriage

   a. is not radically different from marriage
   b. allows for the elimination of traditional gender roles
   c. eliminates many of the problems found among married couples
   d. reduces the level of interpersonal violence between a couple

6. Approximately what percentage of today's young adults (ages 18 to 24) are living at home with their parents?

   a. less than one third
   b. more than half
   c. more than three-quarters
   d. less than half

7. There are several disadvantages in returning to the parental home to live. What represents the most common complaint voiced by members of both generations?

   a. increased drug abuse by children
   b. overdependence on parents
   c. lack of privacy
   d. financial problems

8. Which of the following statements is the *least* accurate regarding homosexual and lesbian relationships?

   a. Gays and lesbians are a varied group, reflecting all occupational fields, political outlooks, and racial groups.
   b. "One-night stands" are much less common among lesbians than among gay men.
   c. Lesbians are more often detected and harassed than gay men.
   d. On the whole, homosexual adults resemble heterosexual adults in their physical health and feelings of happiness or unhappiness.

9. Researchers tend to disagree with psychoanalyst Irving Bieber's theory that male homosexuals come from families in which the

   a. mother is absent
   b. father is dominant
   c. mother is weak
   d. father is ineffective

211

10. Administering male hormones to male homosexuals

   a. alters their sexual performance
   b. increases their sex drive
   c. induces them to display more masculine interests
   d. only increases their gender nonconformity

11. Polyandry is a form of marriage characterized by

   a. two or more husbands and two or more wives
   b. one husband and two or more wives
   c. one wife and two or more husbands
   d. one wife and one husband

12. Currently, approximately what percentage of marriages end in divorce in the United States?

   a. 4 out of 10
   b. 9 out of 10
   c. 1 out of 10
   d. 7 out of 10

13. In sociologist Reuben Hill's nine stages in the family life cycle, the period from departure of the first child to departure of the last child is termed the stage of the

   a. school-age family
   b. aging family
   c. preschool family
   d. family as a launching center

14. Recent research on the health of the American family indicates that

   a. we are returning to the more orderly family life we had prior to 1950
   b. women are adopting vocational patterns of secular celibacy
   c. most parents do not want their children to adopt their lifestyles
   d. most people are pleased with the way their family members rally behind each other in crisis

15. The accomplishment of the developmental tasks accompanying pregnancy is expedited by the woman

   a. taking childbirth training classes
   b. fusing her identity with that of the fetus
   c. defining herself as a parent
   d. having a husband pledged to share

16. Researchers have found that having a first baby may

   a. lead to an increased divorce rate
   b. have a stabilizing effect on marriages
   c. lead to increased romantic intimacy among married couples
   d. cause fathers to become more heavily immersed in their jobs

17. If a child has a working mother, he or she

   a. tends to have less traditional sex-role concepts
   b. is more likely to have retarded cognitive development
   c. is more positively affected if the mother works because of financial necessity
   d. is more likely to become a delinquent

212

18. Which of the following explanations has *not* been advanced for the postpartum blues occurring in some new mothers?

    a. hormonal changes associated with childbirth
    b. feelings of being overwhelmed by the responsibility of caring for another human being
    c. psychological depression stemming from a mother's love-hate conflict with herself
    d. psychological depression stemming from a mother's inability to return to her pre-pregnancy weight

19. Which of the following does *not* pose a particular problem for single-parent mothers?

    a. inflexible school hours
    b. social stigma
    c. remarriage possibilities
    d. social support

20. Oscar has inherited enough money from a rich uncle to live comfortably for the rest of his life. Research reveals that Oscar is likely to

    a. quit his job and retire
    b. continue working
    c. buy a small business
    d. buy out the company he works for

Conceptual Questions:

1. You have joined the local PTA at your son's high school because you are concerned about combating the growing drug problems. Your participation in the PTA

    a. is an example of a conventional morality
    b. helps to provide you with primary relationships
    c. is an example of an instrumental tie
    d. is an example of an expressive tie

2. Dwayne, age 25, is single and lives alone. Which of the following predictions about Dwayne's behavior is the *most* likely to be accurate?

    a. He is likely to never marry.
    b. He is not like other single people because he lives alone.
    c. He lives alone to avoid getting married.
    d. He can be placed in the same category with young women and elderly widows.

3. Although two of your sisters got married in their early twenties, you decide to wait until you are finished with graduate school before you tie the knot. It appears as if your decision

    a. was a wise one because you avoided the marriage squeeze
    b. was ill-advised, and you will probably have to forgo marriage
    c. will not hurt you, and your chances for getting married are quite good
    d. was ill-advised because you will now have fewer men to choose from

4. Your daughter has just told you that she's living off-campus with her boyfriend. Your best advice to her would be which of the following?

    a. Show a strong interest in domesticity so that he will be more likely to marry you.
    b. Just remember that even though you're not married, it may seem like you are still tied down.
    c. All he wants is free sex; after all, why buy the cow when the milk is free?
    d. If you impress him and baby him, he will view this as a trial marriage.

5. Joyce is a middle-aged lesbian. Which of the following predictions about Joyce's behavior is *least* likely to be accurate?

    a. In most respects, Joyce appears indistinguishable from the larger heterosexual population.
    b. Joyce is most likely to come from a low-income, minority family.
    c. Joyce is more likely to be psychologically distressed than a heterosexual woman.
    d. The chances of Joyce's developing a durable, affectionate relationship with another woman are quite high.

6. Your brother is a practicing homosexual. Which of the following observations might have tipped you off to this possibility?

    a. Your uncle on your father's side of the family is a homosexual.
    b. Your mother was the dominant parent during you and your brother's childhood.
    c. As a boy, your brother had a minimal relationship with your father.
    d. As a boy, your brother avoided rough-and-tumble activities, preferring instead to play by himself.

7. Aretha and her husband have two sets of twin boys: James and Richard (who are identical twins) and Clyde and Jeremy (who are fraternal twins). According to recent research on homosexuality, which of the following predictions is the most legitimate?

    a. It is more likely that both James and Richard are gay than both Clyde and Jeremy.
    b. The two sets of twins have equally likely chances of both members becoming gay.
    c. If at least one of the twins is gay, then the chances of Aretha having a lesbian daughter are increased.
    d. Clyde and Jeremy have higher levels of testosterone than James and Richard.

8. The phrase that best describes research on the disintegration of the American family would be which of the following?

    a. The early reports of my death were greatly exaggerated.
    b. The family that prays together stays together.
    c. Ring out the old, and bring in the new.
    d. The road to Hell is paved with good intentions.

9. Julianne is pregnant with her first child. Which of the following behaviors is Julianne the *least* likely to exhibit before the birth of her child?

    a. She will grow more and more preoccupied with the fetus inside of her and begin to define herself as a parent-to-be.
    b. She will attempt to reconcile with her own mother, whom she has remained distant from for the past few years.
    c. She will experience some anxiety concerning the loss of independence which accompanies motherhood.
    d. She will view her fetus as a part of herself, and she will be unable to establish a distinct sense of self prior to the child's birth.

10. Sheila and Stewart are about to have their first baby. Which of the following predictions would be the most reasonable upon the birth of their child?

    a. Sheila and Stewart will experience a short-lived, renewed romantic interest in each another.
    b. Sheila will feel more satisfied with the quality of their marital life than Stewart.
    c. Sheila and Stewart will initially show each other less affection and spend less time sharing leisure activities.
    d. The likelihood of Sheila and Stewart divorcing will dramatically increase.

214

11. Shortly after the birth of your first child, you come home and find your wife crying at the kitchen table. We can reasonably infer that she

   a. is adjusting normally to her new role as mother
   b. is displaying a manic-depressive episode
   c. feels upset over her loss of independence
   d. was abused by her mother as a youngster

12. For the first ten years of her marriage, Mary has been a traditional homemaker. Now she wants to return to the work force, but her husband says that the children will be adversely affected by being deprived of their mother. In her defense, Mary's *most* legitimate argument would be which of the following?

   a. If I stay at home frustrated as I am now, the kids will be more adversely affected.
   b. We'll be OK; you and the children can learn to help out more.
   c. Our children are too old to be affected by my returning to work.
   d. My return to work will improve the children's school grades.

13. A judge is trying to decide whether or not he should grant child custody to the father of an 8-year-old girl. The strongest argument against the father might be that

   a. males are less adequate nurturers of children during the tender years and are therefore poor models
   b. as his daughter approaches puberty, he will be less likely to inform her about the sexual changes that will occur in her body
   c. sexual abuse is far more likely to occur if the father gets custody than if the mother does
   d. he will be more inclined to assign his daughter unreasonable household chores

14. Which statement does *not* describe the significance of work to adult Americans?

   a. Work is a matter of status in American society.
   b. Work is something useful to do between getting up and going to bed.
   c. The measure of a man is work.
   d. Every person's wish is to be so rich that he doesn't have to work.

15. Your family doctor has three sons who are also doctors. The most logical explanation for this is that

   a. your doctor and his wife are genetic geniuses
   b. your doctor has great prestige in the community
   c. your doctor emphasized the importance of education and finishing college
   d. your doctor and his wife both grew up very poor and wanted a better life for their sons

Matching: Definitions

Match the letter of the term in the box with its definition.

a.	bisexuality	h.	path analysis
b.	expressive tie	i.	primary relationship
c.	socioeconomic life cycle	j.	social relationship
d.	secondary relationship	k.	marriage
e.	family life cycle	l.	lifestyle
f.	heterosexuality	m.	homosexuality
g.	instrumental tie	n.	postpartum blues
		o.	serial monogamy

1. the sequence of stages that individuals experience over the life span in the course of status attainment _____

2. a statistical procedure employed for determining how status attained at one stage of the life cycle affects the prospects for a subsequent stage _____

3. irritability, frequent crying spells, difficulty sleeping, and other symptoms experienced by some new mothers two or three days after delivery _____

4. a social link formed when we cooperate with another person to achieve a limited goal _____

5. an equal preference for individuals of the same and opposite sex as sexual partners _____

6. a social interaction that rests on an instrumental tie between people and in which individuals need to have little or no knowledge of each other _____

7. a preference for an individual of the opposite sex as a sexual partner _____

8. a social link formed when we invest ourselves in and commit ourselves to another person _____

9. a socially sanctioned union between a woman and a man with the expectation that they will play the roles of husband and wife _____

10. a life pattern of marriage, divorce, and remarriage _____

11. a preference for an individual of the same sex as a sexual partner _____

12. a social interaction that rests on an expressive tie between people and in which the parties experience warmth and closeness _____

13. the overall pattern of living whereby an individual attempts to meet his or her biological, social, or emotional needs _____

14. the sequential changes and realignments that occur in the structure and relationships of a family between the time of marriage and the death of one or both spouses _____

15. an association in which two people become linked together by a relatively stable set of mutual expectations _____

Completion

Supply the term or terms needed to complete each of the following statements.

1. Psychologists suggest that the ability to delay _____ may enhance an individual's occupational status attainment over the life cycle.

2. The _____ thesis depicts the poverty population as a kind of pool with people flowing in and out.

3. Gender _____ is the single most important factor in predicting the eventual sexual preference of homosexual men.

4. For most Americans, the _____ remains a vital, adaptive, resilient human institution.

5. According to the _____ of poverty thesis, economically disadvantaged people in industrial nations lack effective participation and integration within the larger society.

6. The first _____ functions as a major marker or transition and confronts a couple with new developmental tasks.

7. Maternal _____ tends to be associated with children who have less traditional sex-role concepts, who approve of working mothers, and who have a higher evaluation of female competence.

8. In addition to its economic function, _____ helps adults to structure time, and it provides them with a context in which to relate to other people.

9. The social status of a *man's* parents exerts an indirect effect on his _____ attainment because of its effect on a man's level of schooling.

10. Over the past thirty years, _____ families have increased seven times more rapidly than traditional, two-parent, or _____ families.

Multiple-Choice

Factual

1. d (450)
2. a (450)
3. a (451-452)
4. b (452)
5. a (453)
6. b (453)
7. c (454)
8. c (457-458)
9. d (456)
10. b (457)
11. c (458)
12. a (459)
13. d (461)
14. d (463)
15. a (464)
16. b (465)
17. a (467)
18. d (466)
19. c (468-470)
20. b (473)

Conceptual

1. c (450)
2. d (451)
3. c (453)
4. b (453)
5. b (455, 458)
6. d (456)
7. a (457)
8. a (462)
9. d (464)
10. c (465)
11. c (466)
12. b (467)
13. b (472)
14. d (473)
15. c (475)

Matching: Definitions

1. c (474)
2. h (475)
3. n (466)
4. g (450)
5. a (454)
6. d (450)
7. f (454)
8. b (450)
9. k (458)
10. o (459)
11. m (454)
12. i (450)
13. l (450)
14. e (460)
15. j (450)

Completion

1. gratification (475)
2. situational (468)
3. nonconformity (456)
4. family (463)
5. culture (468)
6. child (465)
7. employment (467)
8. work (473)
9. occupational (475)
10. single-parent; nuclear (468)

Chapter 17

Middle Adulthood: Continuity and Change

INTRODUCTION

Chapter 17 focuses on the middle adulthood years. The major emphasis is on the changes, both physical and psychological, which middle-aged individuals experience. The chapter covers three main areas:

- Physical Changes and Health. Some common changes in vision and hearing during middle adulthood are presented. Female mid-life change, including menopause, and male mid-life change are examined in detail. The sexuality and overall health of middle-aged individuals are also discussed.

- Personality Across the Adult Years. Peck's major psychological tasks of middle adulthood are presented, as are three distinct models of personality. Research on personality change and stability is discussed. Continuity and discontinuity in gender characteristics are covered at the end of this section.

- Adaptation Across the Adult Years. Maturity and self-concept are defined and are shown to influence adaptation across the adult years. Research pertaining to the importance of self-esteem and finding happiness in life concludes this chapter.

I. Physical Changes and Health

   A. Vision and Hearing

   B. Menopause and Female Mid-Life Change

   C. Male Mid-Life Change

   D. Sexuality

   E. Health

II. Personality Across the Adult Years

   A. Psychosocial Tasks of Middle Adulthood

   B. Psychological Conceptions of Personality

   C. Personality Continuity and Discontinuity

   D. Dynamic Properties of Growth

   E. Continuity and Discontinuity in Gender Characteristics

III. Adaptation Across the Adult Years

   A. Maturity and Self-Concept

   B. Life Satisfaction

LEARNING OBJECTIVES

After completing Chapter 17, you should be able to:

1. Evaluate the physical and health-related changes that occur in middle adulthood, specifically discussing: (pages 482-486)

   a. vision

   b. hearing

   c. *menopause*

   d. male *climacteric*

   e. female and male mid-life change

2. Categorize the changes in sexual behavior and activity that occur across the adult years. (486-488)

3. Discuss the following health-related problems that occur as adults grow older: (488-494)

   a. heart disease

   b. Type A personality

   c. high blood pressure

   d. cancer

   e. unhealthy habits

4. Characterize the changes in personality that occur across the adult years by discussing the work of Erikson and Peck. (494-495)

5. Compare and contrast the trait, situational, and interactionist models of personality. (495-496)

6. Describe the research data supporting continuity and discontinuity, as well as dynamic views of adult personality development. (496-499)

7. Summarize the research data dealing with the issue of continuity and discontinuity in adult gender characteristics. (499-501)

221

8. Describe the roles played by *maturity*, *self-concept*, and *self-image* in influencing adaptation across the adult years. (501, 503-504)

9. Define *dialectical psychology*, and discuss Riegel's four dimensions of human development. (502-503)

10. Analyze the factors associated with adult life satisfaction. (504-505)

KEY TERMS

---

climacteric (page 483)

crisis (502)

dialectical psychology (502)

maturity (501)

menopause (483)

self-concept (501)

self-image (501)

Multiple-Choice

Circle the letter of the response which *best* completes or answers each of the following statements and questions.

Factual Questions:

1. Which of the following mid-life changes is the most likely to occur?

   a. Recovery from glare occurs more rapidly.
   b. The lens of the eye deteriorates and clouds.
   c. The ability to hear low-pitched notes declines.
   d. The eye loses its ability to "zoom."

2. The end of menstrual activity is known as

   a. androgyny
   b. maturity
   c. menopause
   d. crisis

3. Which of the following statements regarding estrogen-replacement therapy (ERT) is the most accurate?

   a. Estrogen increases the development of osteoporosis.
   b. Estrogen therapy is recommended for all postmenopausal women.
   c. Estrogen therapy is the leading cause of breast cancer in women.
   d. Estrogen therapy is not recommended for women with a family history of breast or uterine cancer.

4. Middle-aged men are likely to experience all of the following *except*

   a. enlargement of the prostate gland
   b. a large decrease in the production of the male sex hormones
   c. increased reflection on their accomplishments and goals
   d. fears about their sexuality and declining physical attractiveness

5. Research by William H. Masters suggests that the greatest obstacle to sexual performance in the later years of life is

   a. self-fulfilling prophecies of sexual inadequacy
   b. the menopause and biological changes associated with it
   c. the mid-life crisis
   d. the degeneration of the genitals

6. Which of the following statements regarding adult sexual activity is the *most* accurate?

   a. Men find it takes longer for them to achieve an erection once they pass into their forties.
   b. Levels of sexual activity in the twenties do not predict later sexual activity levels.
   c. In older men, the use of blood pressure medications may be associated with impotence.
   d. In older women, the walls of the vagina become thicker and are more easily irritated.

7. For people who are 45 and over, the leading cause of death is

   a. cardiovascular disease
   b. kidney failure
   c. malignant tumors
   d. metastasized cells

8. All of the following are common Type A behavior characteristics *except*

   a. concentrating on several activities at once
   b. hurrying others who are speaking
   c. lack of ambition
   d. becoming excessively irritated when forced to wait in line

9. According to psychologist Albert Bandura, self-_____ determines an individual's ability to unlearn bad health habits and cultivate good ones.

   a. concept
   b. efficacy
   c. image
   d. awareness

10. Many medical scientists believe that up to 80 percent of all _____ result in part from contamination of the environment by carcinogens.

   a. cancers
   b. cardiovascular problems
   c. benign tumors
   d. kidney failures

11. According to Peck, individuals cultivate greater understanding and compassion when they confront the mid-life task of

   a. mental flexibility vs. mental rigidity
   b. cathectic flexibility vs. cathectic impoverishment
   c. valuing wisdom vs. valuing physical powers
   d. socializing vs. sexualizing in human relationships

12. According to the _____ model of adult personality development, people seek out congenial environments that reinforce their preexisting bents.

   a. situational
   b. interactionist
   c. dynamic
   d. trait

13. Psychologists like Kelly, Woodruff, and Birren have undertaken longitudinal studies investigating personality constancy and change. They find that when individuals are tested in college and then again twenty to twenty-five years later, they reveal

   a. little overall changes in self-descriptions
   b. substantial changes in vocational interests
   c. radical changes in basic values
   d. major personality change

14. The Baltimore Longitudinal Study found age-related differences that occurred across time. A seven-year follow-up revealed that

    a. the masculinity score of the men rose
    b. men in their twenties had decreased the pace of their activity
    c. men over fifty had decreased their pace of activity
    d. there was an increase among men of all ages in scores associated with friendliness

15. Bernice L. Neugarten suggests that middle age typically brings a(n)

    a. concern with one's legacy or psychological survivorship
    b. renunciation of life's losses
    c. awareness of oneself as the bridge between generations
    d. reconciliation with members of one's family

16. David Gutmann has compared individuals in a number of cultures and finds that with advancing age

    a. both men and women tend to move toward more passive mastery
    b. both men and women tend to move toward active mastery
    c. men tend to move toward passive mastery, and women tend to move toward more active mastery
    d. men tend to move toward more active mastery, and women tend to move toward passive mastery

17. According to Gutmann, in later life those people who age successfully

    a. incorporate both male- and female-typed characteristics within a single personality
    b. have the same view of themselves over time
    c. have mental pictures of themselves that are relatively temporary
    d. move from passive mastery to active mastery

18. The capacity of individuals to undergo continual change in order to adapt successfully and cope flexibly with the demands and responsibilities of life *best* defines

    a. maturity
    b. sanity
    c. stability
    d. satisfaction

19. Jeanne Marecek and David Metke studied the relationship between self-esteem and success. They find that individuals with low self-esteem who are certain of their low self-appraisal

    a. improve their performance on success tasks if they are told that their success is due to their own skill
    b. improve their performance on success tasks if they are told that their success is the product of luck
    c. are incapable of improving their outcomes under any circumstances
    d. are the individuals most capable of successful outcomes

20. Which of the following factors is the most closely associated with an adult's life satisfaction?

    a. self-esteem
    b. intelligence
    c. income
    d. sex

Conceptual Questions:

1. In which of the following situations will a 50-year-old driver be the *most* adversely affected?

    a. driving down a dark country road toward a car that is approaching with its bright lights on
    b. trying to judge the distance necessary to stop the car when driving at high speed
    c. trying to carry on a conversation, listen to the radio, and simultaneously drive in rush-hour traffic
    d. trying to accelerate and merge with traffic when entering a busy expressway

2. You are going through menopause and are concerned because your mother told you about unpleasant hot flashes that she experienced during this time. Your gynecologist should advise you that

    a. symptoms like hot flashes occur only if your ovaries are removed
    b. estrogen-replacement therapy can eliminate hot flashes and profuse sweating
    c. women who report hot flashes and flushes are merely hypochondriacs with psychosomatic complaints
    d. most women who go through this perceive the "bark" of menopause to be worse than the "bite"

3. Conrad, who is in his late forties, has been the main breadwinner in his family since he and his younger wife, Claudine, were married. As Conrad focuses on retirement, Claudine has plans to try to further herself along the corporate ladder. We can reasonably assume that

    a. Conrad will have no trouble adapting to competition from his wife in the breadwinner role
    b. Claudine's involvement in her own career will enable Conrad to relax and avoid a mid-life crisis
    c. Conrad and Claudine's marriage may experience a "crisis point"; Conrad wants to spend more time with his family, but Claudine longs for success in the job market
    d. Conrad and Claudine's marriage may be strengthened because of Conrad's renewed sexual interest, which occurs in most middle-aged men

4. You and your wife have been married for thirty years. While she is still interested in having sex regularly, you are increasingly less interested. A sex therapist would probably tell you that

    a. too much sexual interest in people your age is perverse
    b. your loss of interest may be caused by fear of failure rather than lack of physical ability
    c. your loss of interest is being caused by your wife's persistence and sexual aggressiveness
    d. your loss of interest is predictable because people with active sex lives in their twenties have more difficulty becoming aroused in their fifties

5. Juan is 50 years old, and there is a history of coronary artery disease in his family. The initial advice his doctor might give him to prolong his life expectancy might be to

    a. receive injections of estrogen
    b. have coronary bypass surgery
    c. exercise more and eat fewer fatty foods
    d. limit his exposure to carcinogens

6. An appropriate diagnostic description of the Type A behavior pattern might be

    a. hurry sickness
    b. anxiety disorder
    c. attention-deficit disorder
    d. antisocial personality

7. Which of the following individuals is the *least* likely to die of heart disease?

   a. Mark, a competitive 45-year-old businessman who smokes and drinks regularly
   b. Joshua, a 50-year-old Mormon who watches his diet, exercises regularly, and finds time to relax
   c. Julianne, an aggressive 42-year-old stockbroker who avoids eating red meat to monitor her high cholesterol
   d. Tanya, a 45-year-old fitness instructor with a family history of high blood pressure

8. Victoria has gained 25 pounds over the past few years. Her physician has recently advised her that she must try to shed this extra weight for health reasons. Which of the following factors will *not* aid Victoria in her weight loss?

   a. a close friend who has recently lost weight lends his support
   b. a support group such as Weight Watchers
   c. Victoria's husband, who is critical of his wife's weight problem
   d. Victoria's own recognition of the severity of the weight problem and her concern to be more attractive

9. A 45-year-old woman decides to get a silicone breast implant to please her younger boyfriend, just as she pleased him by changing her wardrobe. Robert Peck would view this behavior as a form of

   a. generativity
   b. stagnation
   c. mental rigidity
   d. valuing physical powers

10. John acts very lively and extroverted when he is with his friends but very sedate and conservative at work. This observation is the most consistent with which of the following models of personality?

   a. Michel's situational model
   b. the dynamic model
   c. the trait model
   d. the interactionist model

11. Extroverted adults tend to be drawn toward social situations in which they can meet new people, socialize, and enjoy themselves; whereas they avoid situations where people act aloof, clannish, and antisocial. This observation would be consistent with the _____ model of adult personality.

   a. dynamic
   b. interactionist
   c. situational
   d. trait

12. A woman seemingly changes her personality style during menopause, becoming more cold, anxious, and depressed. Which of the following conclusions seems to be the *most* legitimate?

   a. Menopause is an inner biological event that is commonly associated with major personality change.
   b. The woman's husband experienced similar personality changes when he went through male climacteric.
   c. The woman possessed all these personality predispositions earlier in adult life.
   d. The woman has probably displayed major changes in intellectual and cognitive functioning, as well.

13. Which of the following descriptions seems to best characterize the dynamic properties of personality during the middle adult period?

    a. "Let's hold on to what we've got"
    b. "For the good times"
    c. "Torn between two lovers"
    d. "Take it to the limit"

14. Younger male is to _____ as older female is to _____.

    a. androgyny; passive mastery
    b. passive mastery; passive mastery
    c. active mastery; androgyny
    d. active mastery; passive mastery

15. Which of the following age-related occurrences would be considered *most* likely?

    a. A 40-year-old man gives up his ambition to get a job promotion.
    b. A 60-year-old man who has just retired starts helping his wife with the dishes for the first time.
    c. A 20-year-old man starts thinking about how long he has to live.
    d. A 60-year-old woman becomes more tender and affiliative and starts baking again for the first time in thirty years.

16. When Bill and Sue met, nobody thought that they would get along because he was outgoing and aggressive, and she was quiet and introverted. The differences that stimulated their romance have now resulted in divorce. Which of the following theories would consider this crisis an opportunity for personal growth in Bill and Sue?

    a. dialectical
    b. trait
    c. discontinuity
    d. socialization

17. The dialectical theory of development is to _____ as Piaget's theory of development is to _____.

    a. change; balance
    b. continuity; discontinuity
    c. harmony; change
    d. change; crisis

18. Ann always makes critical remarks about her intelligence. She is most likely to accept a high grade on a term paper if she

    a. has failed on all of her other class projects
    b. thinks she got lucky
    c. thinks she worked hard on the paper
    d. thinks the paper reflects her true ability

19. Which of the following individuals would you expect to report the greatest life satisfaction?

    a. a person who just won the lottery
    b. a woman who was recently divorced after being in an abusive situation
    c. a recently married 25-year-old woman
    d. an affluent, suburban, white male

228

20. Adrienne is a very happy, contented woman. Which factor is the *least* likely to have contributed to Adrienne's happiness?

    a. high self-esteem
    b. a healthy marriage
    c. a fulfilling career
    d. a large inheritance

Matching: Definitions

Match the letter of the term in the box with its definition.

a. maturity	i. self-image
b. androgyny	j. dialectical psychology
c. climacteric	k. atherosclerosis
d. osteoporosis	l. self-concept
e. menopause	m. trait model of personality
f. malignant tumor	n. benign tumor
g. crisis	o. interactionist model of personality
h. glaucoma	

1. a noncancerous tumor in which cells do not multiply or spread uncontrollably _____

2. states that individuals gradually form certain characteristics that become progressively resistant to change with the passage of time _____

3. a disorder generally affecting postmenopausal women in which the bone tissue thins and becomes more vulnerable to fractures _____

4. the mental picture we have of ourselves, which is subject to change as we move from one social situation to another _____

5. a theoretical approach that seeks to understand the changing individual in a changing world; it stresses the contradictions and conflicts characterizing human life _____

6. a thickening of the lining of the blood vessels that is thought to be associated with fatty substances in the blood _____

7. incorporation of both male- and female-typed characteristics within a single personality _____

8. the gradual degeneration of the ovaries and the various biological processes associated with ovarian function _____

9. the view we have of ourselves over time as the "real me" or "I myself as I really am" _____

10. the capacity of individuals to undergo continual change in order to adapt successfully and cope flexibly with the demands and responsibilities of life _____

11. an eye disorder in which pressure builds up inside the eye because of a disturbance in the normal drainage of fluid _____

12. a cancerous tumor that keeps growing and invades neighboring tissues _____

13. the cessation of menstrual activity; one of the most readily identifiable signs of the climacteric in women _____

14. a highly demanding situation in which individuals must adjust their behavior to new sets of circumstances _____

15. states that behavior is always a joint product of the person and the situation _____

Completion

Supply the term or terms needed to complete each of the following statements.

1. Estrogen therapy may be beneficial in the treatment of _____.

2. The _____ model views a person's behavior as the outcome of the characteristics of the situation in which the person is momentarily located.

3. _____ is the psychologist who defined four psychological tasks of middle adulthood.

4. Bernice Neugarten described middle-aged men as _____ who saw themselves as possessing energy congruent with the opportunities presented to them.

5. For many older people, believing that they will lose their sexual effectiveness becomes a self-fulfilling _____.

6. According to Albert Bandura, a strong sense of _____ helps individuals to reach goals and master challenges.

7. Dialectical psychologists hypothesize that conflict between the different dimensions of development leads to crises and to developmental _____.

8. _____ people report considerably less happiness than those who have never married.

9. The American male is often stereotyped in the popular media as preoccupied with _____.

10. Not all individuals with low self-esteem undertake to avoid success, only those who harbor a chronically low _____ _____ and avoid success.

230

Multiple-Choice

Factual

1. d (482)
2. c (483)
3. d (483)
4. b (486)
5. a (486)
6. c (487)
7. a (489)
8. c (491)
9. b (492)
10. a (493)
11. d (494)
12. b (496)
13. a (497)
14. c (498)
15. c (499)
16. c (500)
17. a (500)
18. a (501)
19. b (503-504)
20. a (504-505)

Conceptual

1. a (482)
2. b (483)
3. c (486)
4. b (486)
5. c (488-490)
6. a (491)
7. b (488-490)
8. c (492-493)
9. d (494)
10. a (495-496)
11. b (496)
12. c (497-498)
13. d (498-499)
14. b (499-500)
15. b (500)
16. a (502)
17. a (502)
18. b (504)
19. c (505)
20. d (505)

Matching: Definitions

1. n (492)
2. m (495)
3. d (483)
4. i (501)
5. j (502)
6. k (489)
7. b (500)
8. c (483)
9. l (501)
10. a (501)
11. h (482)
12. f (492)
13. e (483)
14. g (502)
15. o (496)

Completion

1. osteoporosis (483)
2. situational (495)
3. Peck (494)
4. risk takers (498)
5. prophecy (486)
6. self-efficacy (492)
7. change (502)
8. divorced (505)
9. sex (487)
10. self-concept (504)

Chapter 18

Middle Adulthood: Love and Work

INTRODUCTION

Chapter 18 focuses on some of the changing aspects of mid-life associated with the family life cycle and the workplace. Two main topics are addressed:

- The Psychosocial Domain. The importance of developing friendships and becoming part of a couple during adulthood are examined. Issues such as extramarital sexual relationships, divorce, and remarriage are all covered in detail. Some difficulties as well as opportunities encountered by stepfamilies as a result of remarriage are considered. The empty-nest period-- when children have grown up and left home--is discussed, along with the topic of adults caring for their elderly parents.

- The Workplace. Individual satisfaction derived from work and the factors associated with job burnout are discussed. Mid-life career changes are also examined. The impact of unemployment and the four stages commonly resulting from job loss are outlined. Then the author addresses female work force expansion and its effects on women and traditional gender roles. Finally, the significance of dual-career couples is presented.

CHAPTER OUTLINE

I. The Psychosocial Domain

    A. Social Behavior

    B. Couples

    C. Extramarital Sexual Relations

    D. Divorce

    E. Remarriage

    F. Stepfamilies

    G. The Empty Nest: Postparental Life

    H. Care for Elderly Parents

II. The Workplace

    A. Job Satisfaction

    B. Mid-Life Career Change

    C. The Impact of Unemployment

    D. Gender Roles: Persistence and Change

    E. Dual-Earner Couples

LEARNING OBJECTIVES

After completing Chapter 18, you should be able to:

1. Describe friendship and couple patterns in the United States. (pages 510-512)

2. Briefly discuss the topic of extramarital sex. (512-514)

3. List the difficulties adults encounter with divorce and uncoupling. (514-516)

4. Discuss the issue of remarriage and the problems and opportunities encountered by stepfamilies. (516-522)

5. Develop an opinion regarding the issue of battered women. (518-519)

6. Demonstrate the significance of the following phenomena: (522-525)

   a. the *empty nest syndrome*

   b. being a member of the *sandwich generation*

   c. *religious practices*

7. List the factors associated with job satisfaction and *job burnout*. (525-528)

8. Explain the significance of mid-life career change. (528-529)

9. Identify the reactions to, and effects of, unemployment in adulthood. (529-530)

10. Describe some of the effects that expansion of the female work force has had on women and traditional sex roles. (531-534)

11. Appraise the significance of dual-career couples. (534-535)

KEY TERMS

---

alienation (page 525)

displaced homemaker (516)

empty nest (522)

empty nest syndrome (522)

job burnout (527)

religion (525)

sandwich generation (523)

235

Multiple-Choice

Circle the letter of the response which *best* completes or answers each of the following statements and questions.

Factual Questions:

1. Which of the following statements regarding friendship is *least* accurate?

   a. Most Americans have only one or two really close friends.
   b. The quality of men's and women's friendships is very similar.
   c. Many men have difficulty making friends on the job.
   d. Men typically talk with friends about job-related issues, sports, and politics and spend little time talking about feelings.

2. Blumstein and Schwartz, in investigating the experiences of different types of couples, found that

   a. straight and homosexual women emphasize power and dominance in their relationships
   b. even when wives had full-time jobs, they did most of the housework
   c. married couples measured their financial success in terms of total earned income
   d. quantity and quality of sex were more important for homosexual couples than for heterosexual couples

3. Which of the following statements regarding sexual relations is accurate?

   a. Americans seem to be giving up affairs entirely because of the AIDS threat.
   b. Younger women are as likely as their male peers to enter extramarital relations.
   c. Extramarital transgressions of the husband usually indicate dissatisfaction with the marital relationship.
   d. The incidence of marital infidelity has declined as couples realize that sexual activity is their marital obligation.

4. Wives are more likely than husbands to

   a. engage in extramarital activity
   b. engage in extramarital relations because of their parents' transgressions
   c. commit suicide if they get divorced
   d. anticipate the breakup of their marriage

5. The displaced homemaker is likely to be devastated by divorce because

   a. divorce is more uncommon in this group of people
   b. she has a more difficult time forming new sexual relationships
   c. she is ill-equipped to deal with the financial consequences of divorce
   d. she is more likely than the working wife to have positive feelings toward her husband

6. A woman who has been physically abused by her husband is

   a. more likely to stay with him if she has fewer resources and little power
   b. likely to seek outside help if the abuse occurs regularly
   c. less likely to remain with him if she was also abused as a child
   d. more likely to acquiesce to the abuse if she has a job

7. Boys who live in families with a stepfather

   a. show less maturity and confidence than boys from single-parent families
   b. benefit from the introduction of a man into a single-parent household
   c. have less adequate development than boys who live in families with a stepmother
   d. approach their mothers' remarriage more positively than girls do

8. For members of the sandwich generation

   a. responsibility for the elderly falls most commonly on daughters
   b. men are most likely to experience role overload
   c. elderly parents are considered a burden and are rarely taken care of
   d. elderly parents are more likely to be taken care of if the parents live far away

9. One of the major functions of religion is to

   a. help women cope with the empty-nest syndrome
   b. create dependency on a supreme being
   c. increase marital happiness and enjoyment
   d. help people deal with the breaking points of life

10. The best defenses against job burnout are

    a. good pay and job security
    b. apathy and fatalism
    c. self-insight and self-awareness
    d. idealism and high expectation

11. Several hypotheses have been advanced to explain why older people are more satisfied with their jobs than younger people. Proponents of the "life-cycle effects" hypothesis say that

    a. the "now" generation of workers holds values that contradict those of the industrial order in which they work
    b. older workers are more satisfied with their jobs because of aging effects
    c. older workers are more satisfied with their jobs because they have better jobs than younger workers
    d. there has been a weakening of materialistic standards, so younger workers are less responsive to traditional rewards

12. According to a recent survey, what percentage of Americans polled switched careers at least once?

    a. more than three-quarters
    b. over half
    c. less than one-third
    d. one-quarter

13. Since 1990, sex labeling or the segregation of occupations by sex in the United States has

    a. changed greatly
    b. not changed
    c. changed little
    d. changed moderately

237

14. Women have been less likely to reach senior-level management positions because

    a. men have traditionally felt that women would not fit in, get along, or feel comfortable in a leadership role
    b. women receive more negative feedback from their superiors
    c. women have too many male mentors
    d. women work primarily for pin money

15. One major source of tension in dual-career families is

    a. disagreements over how the extra money should be spent
    b. the attempt by the husband to eliminate traditional gender roles
    c. the husband's ability to earn more money than the wife
    d. the difficulty of scheduling time together

Conceptual Questions:

1. Which of the following scenarios is the *most* likely to occur?

    a. Shirley and Terrence are a married couple in which Terrence is happiest when he does the lion's share of the housework.
    b. Sam and Diane are a working, married couple who measure their financial success in terms of their collective net income.
    c. Anthony and Erica are newlyweds whose primary conflict involves Erica's contention that Anthony spends too much time with his buddies.
    d. Naomi and Rupert have been married for twenty years, and they argue frequently about Rupert's demands to spend more time with his buddies.

2. Which of the following social interactions is the *most* likely to occur?

    a. A middle-aged man tells his best friend, "I am concerned about being impotent."
    b. A middle-aged man spends most of his time playing cards with his buddies.
    c. A woman tells her husband that he can send out the invitations to the family reunion.
    d. A man continues playing basketball after sustaining an injury because his buddies need him to win the game.

3. Kristine has been living with Jose for over two years. Recently she has decided that she wants out of the relationship. Which of the following actions will *not* aid Jose in sensing Kristine's dissatisfaction?

    a. Kristine has taken several vacations alone in the past few months.
    b. Kristine has recently purchased a variety of alternative-lifestyle magazines.
    c. Kristine complains often about Jose when they go out together in public.
    d. Kristine has begun seeking friendships with a number of Jose's friends.

4. A 45-year-old adult is being treated by a psychiatrist because of clinical signs of depression (e.g., insomnia, agitation, feelings of worthlessness). This adult is *most* likely to be a

    a. recently divorced displaced homemaker
    b. divorced mother who initiated the divorce
    c. recently divorced working mother with two children in elementary school
    d. soon-to-be-divorced woman who is having her third affair within a single year

5. Jim is a 45-year-old divorced man. Without knowing anything else about Jim, we are likely to predict that he

    a. is not likely to remarry
    b. will remarry an older woman
    c. will remarry and will have a good chance of getting divorced again
    d. is less likely to remarry than a divorced woman of the same age group

6. Jill has been married to an abusive husband for five years. Whenever he goes out drinking, he beats her up when he comes home. Jill is likely to stay in this situation if she

    a. gets herself a job
    b. is unemployed
    c. has nontraditional values
    d. has a positive self-concept

7. Which of the following scenarios is most likely to take place in a stepfamily?

    a. The wife's son begins to do less well at school after the remarriage.
    b. The wife's son starts behaving in an infantile, regressive manner after the remarriage.
    c. On announcing her plans to remarry, the wife is confronted by her daughter who asks, "Why can't we go back to live with daddy?"
    d. The children do the dishes when their new stepfather asks them to because they know this will make their mother happy.

8. Sarah's youngest daughter has just gotten remarried. Now all her children are gone from the house. We can predict that Sarah will

    a. be depressed as a result of role confusion
    b. have so much new "free time" that she won't know what to do with it
    c. become immobilized because of feelings of loneliness
    d. feel closer to her husband than when they were first married

9. Natasha is 60 years old. While she is still ambulatory, she suffers from severe bouts of rheumatoid arthritis. Which of the following predictions can we make about Natasha?

    a. Natasha's daughter sees her mother regularly and helps her with her chores.
    b. Natasha is likely to reside in a nursing home.
    c. Natasha is likely to be living with her married son.
    d. Natasha's son drives 50 miles a day to see his mother, although his own health is failing.

10. Which of the following individuals would be the *most* satisfied with his job?

    a. a 20-year-old assembly line riveter in an automobile factory
    b. a 50-year-old chef who gets to determine the menu selections each day
    c. a 50-year-old man who has been an assistant vice president at the local bank for ten years
    d. a 20-five-year-old postal clerk who works in the sorting room with 20 other clerks

11. Which of the following employees is *least* likely to change careers?

    a. a nurse who feels that the hospital patients do not appreciate the care she provides for them
    b. an elementary-school teacher who feels that instead of teaching children he is expected to be their surrogate parent
    c. a psychology professor who feels that his students need remedial help and are unprepared to master the material in her courses
    d. a college administrator who has just helped the college eliminate its deficit and is now being asked to computerize its accounting system

12. Jack is a 43-year-old computer engineer who has become disenchanted with his current profession. The enthusiasm has faded, and he lacks a sense of fulfillment. A psychologist would most likely recommend that Jack

   a. continue with his current job because these types of feelings are normal
   b. consider switching careers because it is common for middle-aged adults to reassess what they are doing with their lives
   c. consider getting psychological help because he is on the verge of a nervous breakdown
   d. retire and channel his energies into caring for his family

13. Harold's wife just told him to go out and get a haircut. She also said, "While you're at it, why don't you apply for some jobs, too." The *most* reasonable conclusion one might reach is that Harold

   a. has just received a layoff notice from his employer
   b. has been unemployed for about a month
   c. feels emasculated because he has been unemployed for a long time
   d. will tell his wife, "It's time for you to wear the pants in this family. My turn is up!"

14. Jill is about to graduate from high school. On the basis of the research on female vocational development, her counselor tells her that

   a. if she can afford to stay at home and not work, she should
   b. society will consider her work to be that of a homemaker and mother
   c. she can expect to earn just as much as a man if she gets a college degree
   d. she can make it in the male-dominated job world, but sexism and pregnancy may conspire against her

15. Which of the following dual-career couples is the *least* likely to experience marital difficulties as a result of their careers?

   a. Jeanne, who is a high-paid certified public accountant, and her husband Roland, who is a music teacher
   b. Denise, a full-time nurse who assumes 90 percent of the household chores as well, and her husband Ross, a professor
   c. Lin, a teacher who maintains a separate banking account from her husband Leandro, a physician who shares household duties with his wife
   d. Clarissa, an advertising executive who expects all housework to be done by her husband Bruce, an elementary school teacher

Matching: Definitions

Match the letter of the term in the box with its definition.

a.	alienation	h.	displaced homemaker
b.	empty nest	i.	religion
c.	dual-earner	j.	friendships
d.	job burnout	k.	empty nest syndrome
e.	divorce	l.	discipline
f.	unemployment	m.	sandwich generation
g.	senior-level management positions	n.	stepparents
		o.	sexual harassment

1. serve as vehicles for self-affirmation _____

2. may exact a greater emotional and physical toll than almost any other life stress _____

240

3. women in mid and late life who are often particularly devastated by divorce _____

4. perhaps the biggest source of friction for stepfamilies _____

5. probably the most overlooked group of parents in the United States _____

6. that period of life when children have grown up and left home _____

7. a pervasive sense of powerlessness, meaninglessness, isolation, and self-estrangement _____

8. may have adverse effects on mental and physical health and lead to family life deterioration _____

9. the emotional difficulties that individuals face when their children leave home _____

10. area in which very few women are employed _____

11. the manner in which people deal with the ultimate problems of life _____

12. "unwelcome" sexual attention that affects an employee's job conditions or creates a "hostile" working environment _____

13. a condition that results when work that was once fulfilling and satisfying becomes, over time, unfulfilling and unsatisfying _____

14. most of these families entail junior-senior relationships, with the wife usually secondary in most respects _____

15. middle-aged adults who find themselves with responsibilities for their own teenage children on the one side and for their elderly parents on the other _____

Completion

Supply the term or terms needed to complete each of the following statements.

1. Men's relationships with other men tend to be competitive or activity oriented rather than _____.

2. Research indicates that when _____ did become involved in an extramarital affair, it was more likely to blossom into a full-fledged love affair.

3. Studies indicate that the person who wants out of a relationship, the _____, begins a psychological distancing before the partner realizes that something is wrong.

4. _____ report significantly less satisfaction with their family life than married couples with biological children.

5. Middle-aged women of the sandwich generation are subjected to _____-_____ stresses similar to those experience by younger women in relation to work, child care, and other responsibilities.

6. At times, the realities confronting workers in the _____ professions crush their humanism and make them particularly susceptible to job burnout.

7. Boredom, lack of fulfillment, and disenchantment with one's existing work are often cited as reasons for midlife _____ change.

8. Workers' initial reactions to _____ are expressed through a sequence of shock, relief, and relaxation.

9. A cultural revolution of attitudes and values is required to eradicate the _____ of women.

10. A number of economists note that women who have _____ encounter a substantial career disadvantage.

ANSWERS FOR SELF-TESTS - CHAPTER 18

Multiple-Choice

Factual

1.	b	(510)
2.	b	(511)
3.	b	(513)
4.	d	(516)
5.	c	(516)
6.	a	(519)
7.	b	(519-520)
8.	a	(523)
9.	d	(525)
10.	c	(527)
11.	c	(528)
12.	b	(528)
13.	c	(531)
14.	a	(532)
15.	d	(535)

Conceptual

1.	c	(511)
2.	d	(512)
3.	d	(515)
4.	a	(516)
5.	c	(517)
6.	b	(519)
7.	c	(519-520)
8.	d	(522)
9.	a	(523-524)
10.	b	(526)
11.	d	(527)
12.	b	(528)
13.	c	(530)
14.	d	(533)
15.	c	(535)

Matching: Definitions

1.	j	(512)
2.	e	(514)
3.	h	(516)
4.	l	(521)
5.	n	(518)
6.	b	(522)
7.	a	(525)
8.	f	(529-530)

9.	k	(522)
10.	g	(532)
11.	i	(525)
12.	o	(533)
13.	d	(527)
14.	c	(535)
15.	m	(523)

Completion

1.	intimate (512)		6.	helping(527)
2.	women (514)		7.	career (528)
3.	initiator (515)		8.	unemployment (529)
4.	stepparents (521)		9.	abuse (519)
5.	role-overload (523-524)		10.	children (533)

Chapter 19

Later Adulthood: Physical and Cognitive Development

INTRODUCTION

Chapter 19 focuses on physical and cognitive development during later adulthood. Three topics are looked at in detail, including:

- Aging: Myth and Reality. The age at which an individual is considered elderly differs from one society to another. The demographics of the elderly are presented for various cultures. Why women live longer than men is discussed. A number of myths which cloud the real facts of aging are also outlined, and the overall health of elderly people is covered.

- Biological Aging. Physical changes associated with aging are presented, and a variety of biological theories of aging are examined in this section.

- Cognitive Functioning. The first issue covered in this section is the varied course of different abilities. Results found by studies of late-adult cognitive development and an overestimation of the effects of aging are presented. Research dealing with the memory processes, information processing, and learning in the elderly is also documented. Finally, the topics of senility and Alzheimer's disease are addressed.

CHAPTER OUTLINE

I.   Aging: Myth and Reality

    A.   The Elderly: Who Are They?

    B.   Women Live Longer than Men

    C.   Myths

    D.   Health

II. Biological Aging

    A.   Physical Changes

    B.   Biological Theories of Aging

III. Cognitive Functioning

    A.   The Varied Course of Different Abilities

    B.   Overestimating the Effects of Aging

    C.   Memory and Aging

    D.   Learning and Aging

    E.   Alzheimer's Disease

LEARNING OBJECTIVES

After completing Chapter 19, you should be able to:

1.   Define the terms *gerontology* and *geriatrics*, and describe the demographic statistics on elderly Americans and on the elderly in other cultures. (pages 540-542)

2.   Briefly summarize the research data concerning why women are living longer than men. (542-543)

3. Appraise the concept of ageism by evaluating some common myths concerning elderly Americans. (543-547)

4. Describe the impact that increasing numbers of elderly Americans have on generational tensions. (544-545)

5. Describe the health status of older Americans. (517-530)

6. Discuss the physical changes associated with *senescence*. (550-552)

7. List the biological theories of aging. (552-553)

8. Trace the course of the following abilities in late adulthood: (554-555)

   a. *fluid intelligence*

   b. *crystallized intelligence*

9. Analyze the differing results produced by cross-sectional and longitudinal studies of late-adult cognitive development. (556)

10. Summarize the research dealing with memory processes, information processing, and learning in the elderly. (556-558)

11. Explain the different types of senility; and discuss their causes, the difficulties involved in diagnosis, and the consequences for the victims and their families. (558-561)

12. List several guidelines for caring for a family member with *Alzeheimer's disease*. (560)

KEY TERMS

Alzheimer's disease (page 559)

collagen (550)

crystallized intelligence (555)

death-drop (556)

decay theory (557)

encoding (557)

fluid intelligence (555)

geriatrics (540)

gerontology (540)

hypothermia (551)

interference theory (557)

multiinfarct (558)

osteoporosis (549)

retrieval (557)

senescence (550)

storage (557)

trace (557)

SELF-TESTS - CHAPTER 19

Multiple-Choice

Circle the letter of the response which *best* completes or answers each of the following statements and questions.

Factual Questions:

1. One of the effects of the increasing proportion of elderly people in the United States has been

   a. a lengthening of the life span
   b. an increase in negative attitudes toward the elderly
   c. an increased dependency ratio
   d. a deteriorating standard of living

2. Women seem to be more durable organisms than men because

   a. of an inherited sex-linked resistance to some types of life-threatening disease
   b. lower levels of late-life estrogen protect them against cardiovascular disease
   c. they eat better balanced diets that are lower in cholesterol
   d. they lead more active and energetic lives

3. Which of the following statements is a fact?

   a. Most old people are in poor health.
   b. Less than one-tenth of the elderly who live at home are bedridden and homebound.
   c. Most Americans over 65 find themselves in serious financial straits.
   d. Most of the elderly view crime as a major concern.

4. Which of the following statements regarding the Social Security system is the *least* accurate?

   a. One answer to the problems of the Social Security system has been raising the retirement age.
   b. Today's younger workers must pay Social Security taxes of several thousand dollars a year for their entire working careers.
   c. Critics of the system contend that most people could do much better if they were able to put their Social Security contributions into a private pension plan.
   d. Economists say that we should make Social Security voluntary.

5. When asked to compare their health with that of others their own age, the vast majority of the elderly rate their health as

   a. a serious handicap
   b. good or excellent
   c. bad or poor
   d. slightly below average

6. Elderly women are more likely than men to

   a. experience brief blackouts
   b. absorb drugs rapidly from their intestinal tracts
   c. need calcium supplements to slow or stop bone loss
   d. spend a large percentage of their lives without chronic infirmities

7. The protein substance that has been implicated in the body's aging process is called

   a. dilantin
   b. primarin
   c. collagen
   d. carotene

8. Which of the following is an accurate description of age-related physical change?

   a. a marked loss of hearing, especially in the lower frequencies, in the elderly
   b. gains in maximum oxygen intake
   c. sleep characterized by less frequent awakenings
   d. a decline in the number of taste buds per papilla

9. Which of the following theories of aging implies that the probability of a random happening increases as the number of events increases?

   a. autoimmune mechanism
   b. genetic preprogramming
   c. stochastic processes
   d. error in DNA

10. Tests of some aspects of intelligence, particularly crystallized ability, reveal that with an increase in age, individuals tend to exhibit a(n)

    a. increase in intelligence
    b. substantial decline in intelligence
    c. moderate decline in intelligence
    d. plateau or leveling-off effect

11. Different procedures for studying the effects of aging provide somewhat different results. Which of the following approaches tends to overestimate the effect of aging?

    a. cross-sectional
    b. longitudinal
    c. statistical
    d. case study

12. Research suggests that with respect to aging and memory

    a. a progressive loss of memory necessarily accompanies advancing age
    b. some memory loss is found in an increasing proportion of older people with each advance in chronological age
    c. age-related decreases are more severe for recognition than for recall tasks
    d. there is no clear pattern of problems with memory retrieval

13. According to the decay theory of forgetting

    a. memory failure is due to poor encoding
    b. forgetting is due to the deterioration of memory traces in the brain
    c. retrieval of memory cues is affected as the number of items in a category increases
    d. cue overload is responsible for memory failure

14. Which of the following is *not* characteristic of senility?

    a. progressive mental deterioration
    b. memory loss
    c. disorientation to time and place
    d. impaired hearing

15. In the first phase of Alzheimer's disease

    a. the patient experiences multiinfarcts that destroy small areas of brain tissue
    b. individuals forget where things were placed and have difficulty recalling events
    c. there are difficulties in cognitive functioning that cannot be overlooked
    d. hyperthyroidism leads to signs of depression

Conceptual Questions:

1. On the television show *Ripley's Believe It or Not*, a man in the Soviet Republic of Georgia was interviewed and claimed to be 120 years old. We can probably conclude that

    a. he eats Dannon yogurt
    b. he comes from a rural area and is exposed to less air pollution
    c. his claim to be 120 years old is probably an exaggeration
    d. his dietary habits are healthier than those of most Americans

2. Female is to life expectancy as male is to _____.

    a. estrogen
    b. smoking
    c. immune system
    d. sex-linked resistance

3. Your 10-year-old daughter tells you that she understands why old people are so cranky and crotchety all the time. It is because most of them are sick and invalid and cannot take care of themselves. This perception is

    a. very sophisticated for a 10-year-old
    b. somewhat inaccurate but very close to being true
    c. confused because your daughter has paired the wrong causes with the right effect
    d. inaccurate and may inadvertently perpetuate the inferior status of the elderly

4. You feel very uncomfortable because you get in the middle of a conflict at a family reunion. Your elderly grandfather is berating your cousin because your cousin has been bad-mouthing the Social Security system and the amount of money he has to pay into it. Your cousin's best defense might be which of the following?

    a. Even though you paid your dues, you're getting ripped off.
    b. Your Social Security isn't providing for all your needs, is it?
    c. Pay me now, or pay me later.
    d. You got more for your money than I will.

5. At her last medical checkup, your elderly mother's doctor told her that she had borderline insulin elevation. According to the results of the Baltimore Longitudinal Study, which of the following conclusions would be the *least* reasonable for her doctor to reach?

    a. This medical complication is quite common in elderly people.
    b. Insulin production becomes more efficient in the elderly.
    c. A close examination of her family medical history is in order.
    d. She should immediately be given daily doses of oral insulin.

6. Your grandfather is in the hospital because he fell down and broke his hip. Which of the following problems should you be the *most* concerned about?

    a. He may receive too much estrogen and therefore develop osteoporosis.
    b. He may develop too much bone mass as a result of calcium therapy and be more likely to break his hip again.
    c. He may develop additional problems as a result of receiving other medications, such as sedatives.
    d. He may leave the hospital suffering from a chronic disease that he contracted while in the hospital.

7. Advertisements for various creams and oils imply that these products can keep women's skin less wrinkled and younger-looking. If these products really do make skin younger looking, they accomplish this by

    a. creating more DNA
    b. preventing collagen fibers from changing
    c. attacking defective cells
    d. creating cellular mutations

8. Many state legislatures have passed laws preventing gas and power companies from disconnecting the natural gas supplies of poor and elderly customers. If this legislation were declared unconstitutional, the elderly would be at risk for

   a. decreased ventilatory volume
   b. decreased organ reserve
   c. life-threatening hypothermia
   d. hypoglycemia

9. A biologist places cells from chick embryos in a cell culture to count and determine the number of times the cells divide and redivide. He finds that there is a finite limit to the number of cell divisions that will occur, thus lending support to which theory of aging?

   a. genetic preprogramming
   b. mean time to failure
   c. DNA error
   d. stochastic processes

10. You are copying a computer program from one disc to another when a power outage shuts down your computer. As a result of this problem, your program is only partially transferred to the new disc. This problem is analogous to which of the following theories of aging?

   a. genetic programming
   b. longevity assurance
   c. DNA error
   d. stochastic processes

11. On which of the following tasks is an elderly person *most* likely to experience declines in performance?

   a. crossword puzzles
   b. playing poker
   c. putting a jigsaw puzzle together
   d. playing checkers

12. A corporation president decides to use scores on an aptitude test to promote his executives. The younger executives are receiving a disproportionate number of promotions because they score higher on the test. A psychologist might advise the president that

   a. his method of testing is unfair because it implies that an older worker is a less intelligent employee
   b. if he tested all his employees each year, the older employees would outperform the younger employees
   c. older employees perform more poorly because they are suffering from chronic illness
   d. although older employees score more poorly, they have had better educations

13. An elderly man has just received a past-due notice on his phone bill. He may have forgotten to pay the bill because he got four other utility bills when he received this one and placed them all in a manila folder labeled "bills." The man's memory failure is due to

   a. lack of encoding
   b. the decay of memory trace
   c. interference from other stored information
   d. retrieval of inaccessible information

14. A psychology professor finds that elderly adults perform more poorly on anagrams than younger adults. However, when elderly adults are given practical problems to solve (e.g., what would you do if your refrigerator was warm inside?), they perform as well as younger adults. We can therefore conclude that

   a. the elderly demonstrate learning competence but not performance
   b. memory is not necessary to solve practical learning problems
   c. the elderly perform well only on familiar, habitual tasks
   d. the second task was more relevant for the elderly, and therefore they were motivated to do better

15. Your elderly mother keeps forgetting where she has put her glasses, checkbook, and keys. You should therefore conclude that

   a. she has suffered a stroke
   b. no definite conclusions are warranted without further information
   c. she has Alzheimer's disease
   d. she is suffering from an irreversible organic brain syndrome

Matching: Definitions

Match the letter of the term in the box with its definition.

a.	Alzheimer's disease	k.	interference theory
b.	multiinfarct	l.	decay theory
c.	geriatrics	m.	fluid intelligence
d.	crystallized intelligence	n.	storage
e.	retrieval	o.	senescence
f.	hypothermia	p.	trace
g.	encoding	q.	gerontology
h.	death-drop	r.	osteoporosis
i.	collagen	s.	kuru
j.	cue overload	t.	longevity assurance theory

1. a progressive degenerative disorder that involves deterioration of brain cells _____

2. a "little stroke," or the rupture of blood vessels in the brain _____

3. the ability to make original adaptations in novel situations _____

4. a substance that constitutes a very high percentage of total body protein and appears to be implicated in the aging process _____

5. a disease caused by a slow-acting virus that bears a resemblance to Alzheimer's disease _____

6. maintains that long-living organisms have more cell-repair genes _____

7. a state of being overwhelmed or engulfed by the excessive number of stimuli confronted in a memory task _____

8. changes that occur in the structure and functioning of the organism through time _____

9. a condition in which body temperature falls more than 4° F and persists for a number of hours _____

10. a marked drop in intelligence that may occur just a short time before a person dies _____

11. the residue of an event that remains in memory after the event has vanished _____

12. a condition associated with excessive loss of bone mass, making the bones susceptible to fracture _____

13. the view that retrieval of a cue becomes less effective as more and new items come to be classed or categorized in terms of it _____

14. the branch of medicine that is concerned with the diseases, debilities, and care of elderly persons _____

15. the ability to reuse earlier adaptations on later occasions _____

16. the process by which information is regathered from memory when needed _____

17. the process by which information is retained in memory until it is needed _____

18. the view that forgetting is due to deterioration in the memory traces of the brain _____

19. the process by which information is put into the memory system _____

20. the field of study that deals with aging and the special problems of the elderly _____

Completion

Supply the term or terms needed to complete each of the following statements.

1. _____ theory assumes that aging occurs because the body's natural defenses against infection begin to attack normal cells.

2. The _____ technique underestimates the decline in intelligence that occurs in the elderly.

3. A deficit in choline acetyltransferase may be a causal factor in _____ disease.

4. _____ women have a substantially lower risk of heart disease than men of comparable ages.

5. Many people do not realize that _____ was never intended to provide for all their needs when they retired or became disabled.

6. Elderly women may require _____ supplements to forestall broken hips and fractured wrists in later life.

7. Biologists have suggested that organisms age because their cells are slowly poisoned or hampered in functioning by the waste products of _____.

8. Often, _____ intelligence shows an increase with age, while _____ intelligence shows a drop with age in later life.

9. Psychologists have pointed out that cross-sectional studies of adult aging do not allow for _____ differences in performance on intelligence tests.

10. _____ is typically characterized by progressive mental deterioration, memory loss, and disorientation regarding time and place.

Multiple-Choice

Factual			Conceptual		
1.	c	(541)	1.	c	(540-541)
2.	a	(543)	2.	b	(543)
3.	b	(543)	3.	d	(543)
4.	d	(544-545)	4.	d	(544)
5.	b	(547)	5.	a	(549)
6.	c	(549)	6.	c	(549-550)
7.	c	(550)	7.	b	(550-551)
8.	d	(551)	8.	c	(551)
9.	c	(553)	9.	a	(552)
10.	a	(555)	10.	c	(552)
11.	a	(556)	11.	c	(555)
12.	b	(556-557)	12.	a	(556)
13.	b	(557)	13.	c	(557)
14.	d	(558)	14.	d	(558)
15.	b	(559)	15.	b	(559)

Matching: Definitions

1.	a	(559)	11.	p	(557)
2.	b	(558)	12.	r	(549)
3.	m	(555)	13.	k	(557)
4.	i	(550)	14.	c	(540)
5.	s	(559-560)	15.	d	(555)
6.	t	(553)	16.	e	(557)
7.	j	(557-558)	17.	n	(557)
8.	o	(550)	18.	l	(557)
9.	f	(551)	19.	g	(557)
10.	h	(556)	20.	q	(540)

Completion

1.	autoimmune (552)	6.	calcium (549)
2.	longitudinal (556)	7.	metabolism (553)
3.	Alzheimer's (560)	8.	crystallized; fluid (555)
4.	premenopausal (543)	9.	generational (556)
5.	Social Security (545)	10.	senility (558)

Chapter 20

Later Adulthood: Psychosocial Development

INTRODUCTION

Chapter 20 takes a close look at the period of late adulthood and the psychosocial domain. Several important topics are covered, including:

- The Psychosocial Domain. The psychosocial tasks of later adulthood are described, as viewed by Erikson and Peck. Major late-adulthood personality patterns are introduced. Research data are also presented from an examination of the relationship between self-concept and life satisfaction in late adulthood.

- Theories of Adjustment to Aging. Four main theories describing changes in the elderly in terms of the changes in their social environment are examined. These include (1) the disengagement theory, (2) the activity theory, (3) the role exit theory, and (4) the social exchange theory.

- The Psychosocial Aspects of Aging. Several topics are covered in this section, beginning with retirement, changing marital relations, and kin and friendship ties. Data dealing with the roles and functions performed by grandparents are then presented. A close examination is made of different types of institutional care for the elderly. Finally, the significance of personal control and choice in influencing the well-being of elderly nursing home residents is discussed.

CHAPTER OUTLINE

I. The Psychosocial Domain

   A. Psychosocial Tasks of Later Adulthood

   B. Personality and Patterns of Aging

   C. Self-Concept and Life Satisfaction

II. Theories of Adjustment to Aging

   A. Disengagement Theory

   B. Activity Theory

   C. Role Exit Theory

   D. Social Exchange Theory

III. Psychosocial Aspects of Aging

   A. Retirement

   B. Marital Relations

   C. Kin and Friendship Ties

   D. Grandparents

   E. Institutional Care

   F. The Impact of Personal Control and Choice

LEARNING OBJECTIVES

After completing this chapter, you should be able to:

1. Describe the views of Erikson and Peck on psychosocial development in late adulthood. (pages 566-567)

2. Name the major late-adulthood personality patterns identified by Bernice Neugarten. (567-568)

3. Explain how self-concept is related to life satisfaction in late adulthood. (568-569)

4. Compare and contrast the following theories of adjustment to aging: (569-572)

   a. *disengagement theory*

   b. *activity theory*

   c. *role exit theory*

   d. *social exchange theory*

5. Discuss how the social life space of adults is affected by the following: (572-580)

   a. retirement

   b. changing marital relations

   c. kin and friendship ties

6. Briefly summarize the research data dealing with: (578-584)

   a. types of home care for the elderly and its consequences

   b. types of institutional care for the elderly and its consequences

7. Cite the data dealing with the following aspects of grandparenting: (580-581)

   a. demographic statistics

   b. roles and functions performed

8. Assess the significance of the following factors in the selection of a nursing home: (582-583)

   a. the facility

   b. the safety of the home

   c. the quality of the staff

   d. the types of activities and foods offered

   e. the atmosphere of the home

   f. the cost of the home

9. Explain the importance of personal control and choice in influencing the well-being of elderly nursing home residents. (584-585)

KEY TERMS

---

activity theory of aging (page 570)

disengagement theory of aging (569)

modernization theory (572)

role exit theory of aging (571)

social exchange theory of aging (571)

Multiple-Choice

Circle the letter of the response which *best* completes or answers each of the following statements and questions.

Factual Questions:

1. An elderly person who sees himself as having multiple dimensions and as pursuing new ways of finding a sense of satisfaction is demonstrating

   a. disengagement
   b. ego differentiation
   c. body transcendence
   d. ego transcendence

2. In Neugarten's study of personality patterns in the aged, she described the disintegrated elderly as

   a. revealing gross defects in psychological functions and an overall deterioration in thought processes
   b. well-functioning individuals who reveal a complex inner life and intact cognitive abilities
   c. striving, ambitious, achievement-oriented individuals with strong defenses against anxiety
   d. having strong dependency needs and eliciting responsiveness from others

3. Older people have more favorable self-concepts if they

   a. live in institutions
   b. are poverty-stricken
   c. are poor in health
   d. are relatively affluent

4. Research suggests that being close to one's siblings while in college

   a. has little effect on well-being in late adulthood
   b. more strongly predicts emotional well-being in late adulthood than having had a successful career
   c. is a weaker predictor of emotional well-being in late adulthood than having had a good marriage
   d. is associated with psychological depression and high divorce rates in late adulthood

5. According to which of the following theories does a gradual and mutually satisfying process occur in the course of aging in which society and the individual prepare in advance for incapacitating disease and death?

   a. role exit
   b. disengagement
   c. social exchange
   d. activity

6. According to activity theory

   a. as people age, they lose their master status
   b. a sharp drop occurs in the involvement of the elderly with voluntary associations
   c. the decreasing social interaction of the elderly results from the withdrawal of society from the aging person
   d. the elderly attempt to extract from society a more favorable distribution of benefits and privileges for themselves

7. The social exchange theory of aging suggests that the elderly find themselves in a state of increasing vulnerability because

   a. they have less to offer society
   b. role loss is a stressful experience
   c. increasing numbers of elderly individuals are retiring early
   d. as an elderly person's activity level declines, so do feelings of satisfaction, contentment, and happiness

8. The _____ theory assumes that the position of the aged in preindustrial, traditional societies is high because the aged tend to accumulate knowledge and control through their years of experience.

   a. activity
   b. disengagement
   c. modernization
   d. social exchange

9. Many Americans view the practice of compelling workers to retire as a

   a. way of ensuring more jobs for young adults
   b. long-term solution to inflation
   c. curtailment of basic rights
   d. way of improving future life prospects for the elderly

10. According to the Bureau of Labor Statistics, by the year 2000 only one in _____ men 60 years of age and over will be working.

   a. two
   b. eight
   c. four
   d. six

11. In regard to recent research on retirement, which of the following statements is the *most* accurate?

   a. More than two-thirds of all Americans find retirement stressful.
   b. It is money that is the most missed in retirement.
   c. Most men who retire are very unhappy and would choose to retire later if they could choose again
   d. On the whole, voluntary and nonvoluntary retirees seem to be equally disenchanted with retirement.

12. Which of the following factors appears to contribute to improved marital relationships in later years?

    a. Problems with issues such as in-laws and sex have often been resolved.
    b. Grown children are more likely to show appreciation and affection toward their elderly parents.
    c. Each partner has more time to pursue personal interests.
    d. Grandchildren are more likely to help out with chores that the couple cannot perform.

13. Which of the following statements regarding kin and friendship of the elderly is the *least* accurate?

    a. Friends are more important and satisfying overall to older persons in terms of companionship than relationships with their offspring.
    b. Many grandmothers resent the frequent imposition that their adult children make on their time and energy, such as babysitting for grandchildren.
    c. Siblings often play a significant role in the lives of the elderly.
    d. Those elderly persons with adult children prefer to live with their offspring than alone.

14. Elderly people who are suffering from some physical disability

    a. eat and sleep less well in familiar surroundings
    b. become more withdrawn and depressed if they live at home
    c. take less responsibility for their own care when they are at home
    d. recover more rapidly in a home than in a hospital environment

15. Which of the following statements concerning the current generation of grandparents is accurate?

    a. More youngsters have living grandparents.
    b. Today's grandparents have less money.
    c. Grandfathers report greater satisfaction with grandparenting than grandmothers.
    d. The most common role for grandparents is that of surrogate parents.

16. Approximately what percent of persons aged 65 and over reside in nursing homes?

    a. 2 percent
    b. 12 percent
    c. 5 percent
    d. 15 percent

17. According to your text, most nursing homes

    a. have an adequate, well-qualified staff
    b. are owned by private proprietors and are operated for profit
    c. are rarely used by terminally ill patients who require intensive nursing care
    d. are more economical than services such as sheltered housing and "meals on wheels"

18. Because of increased government regulations, nursing homes have become

    a. warehouses for the elderly and dying
    b. a better alternative than home care for the elderly
    c. miniature hospitals with bureaucratic requirements
    d. safer and more wholesome environments for the elderly

19. Nursing home residents who were told by an administrator that they were responsible for caring for themselves and for shaping the home's policies

    a. became more depressed than comparable residents
    b. were rated as healthier than comparable residents
    c. were overwhelmed by their responsibilities and refused to assume them
    d. expressed a preference to be physically and emotionally dependent

20. Ellen Langer and Judith Rodin undertook a study in a nursing home and found that the patients who were given a choice and asked to make decisions

    a. fell into psychological depression by virtue of the stress
    b. exhibited helplessness in the hopes of getting the staff to resume decision making
    c. began to take over and order the staff about
    d. showed an improvement in terms of activity, happiness, and sociability

Conceptual Questions:

1. The song title that would best describe Erikson's notion of ego identity would be

    a. "Morning Has Broken"
    b. "Where Is the Love?"
    c. "I Did It My Way"
    d. "Let Me Down Easy"

2. An elderly man tells you, "I live every day as if it were the first day of my life, or it might be the last." People can say that he did the best he could with the tools he had. This statement indicates a satisfactory resolution of

    a. body transcendence vs. body preoccupation
    b. ego differentiation vs. work-role preoccupation
    c. ego transcendence vs. ego preoccupation
    d. generativity vs. stagnation

3. Your grandfather is a retired corporate executive and has now become involved full time in the retired senior volunteer program helping people start new businesses. He claims that he doesn't want to become an old "fuddy-duddy" in a rocking chair. Your grandfather could be characterized as a(n)

    a. reorganizer
    b. disintegrated type
    c. passive-dependent type
    d. armored-defended type

4. Which of the following situations is the *most* likely to occur?

    a. When given a questionnaire measuring life satisfaction, a group of 60-year-old men score lower than a group of 30-year-old men.
    b. When given a questionnaire measuring life satisfaction, a group of 30-year-old men get the same average score as a group of 60-year-old men.
    c. Your grandfather tells you that his retirement years are far worse than he expected them to be and that he is bored.
    d. A group of 30-year-old men express less satisfaction with their life accomplishments than a group of 60-year-old men.

262

5. A 60-year-old member of your local Rotary Club has quit the organization after being a member for thirty years. Now he hardly even associates socially with the men he used to see each day. We can *most* reasonably conclude that

   a. his behavior would be more likely to occur in someone much older than he is
   b. his behavior is a normal way of acknowledging that organizations need "new blood"
   c. as people get older, they disengage from activities in which they no longer feel useful
   d. he resigned because he felt that the club didn't want old members, and he didn't want to be a burden

6. According to the predictions of activity theory, an elderly woman who sits in her apartment and watches television all day

   a. will not be as happy as an elderly woman who keeps herself occupied with a variety of tasks and social activities
   b. does so because she feels that slowing down is an inevitable and desirable result of aging and is preparing herself for the "ultimate disengagement" of incurable, incapacitating disease
   c. is preparing herself for the "ultimate disengagement" of incurable, incapacitating disease
   d. can be happy with reduced activity levels as long as her health is OK

7. Master status is to _____ as role loss is to costs.

   a. core role
   b. profit
   c. social exchange
   d. traditional knowledge

8. Which of the following individuals should have the highest reported life satisfaction?

   a. a 60-year-old man who retired because of chronic emphysema
   b. a college professor who retired because he reached the mandatory retirement age of 70
   c. an automobile worker who retired because he was about to be laid off
   d. a legal secretary who retired when her law firm implemented a better pension plan

9. The *best* geographical way to describe the relationship between length of marriage and marital satisfaction would be to draw a

   a. hill
   b. plateau
   c. series of hills and valleys
   d. valley

10. An elderly person is feeling lonely, isolated, and depressed. Which of the following individuals is the *most* likely to be experiencing these symptoms?

    a. an elderly widow who went to live with her son's family in a distant city
    b. an elderly widower who can be seen playing cards every day with a group of men at the senior citizens center
    c. a widower who takes daily walks with a longtime acquaintance
    d. an elderly lifelong bachelor who lives by himself

11. Your elderly mother fell down on the icy pavement in front of her car and broke her arm. The best course of action would be to

    a. put her in a hospital so that her recuperation will progress more rapidly
    b. hire a full-time housekeeper for her because she will not be capable of caring for herself alone
    c. place her in a residential care facility where her physical and psychological needs can be taken care of
    d. let her recuperate at home and provide her with assistance when and where needed

12. The song title that would best describe the relationship between today's grandparents and their grandchildren would be

    a. "The Pretender"
    b. "Five Short Minutes"
    c. "You've Got a Friend"
    d. "When Time is Stolen"

13. Which of the following elderly persons would be the *least* likely candidate to be placed in a residential nursing home?

    a. a homeless person with no family and nowhere else to go
    b. an elderly victim of Alzheimer's disease who is incontinent
    c. an elderly man who has been paroled after a long imprisonment
    d. an elderly person recovering from major cancer surgery

14. You and your wife are the only living relatives of your disabled and terminally ill elderly mother. However, when attempting to find a suitable nursing home for her, you are most disconcerted when you find that the home

    a. takes only Medicare and Medicaid patients
    b. staff interact with patients only for caretaking purposes
    c. is not federally licensed
    d. has an institutional appearance with dark, drab paint

15. Your mother was alert and talkative when she entered a nursing home. One month later she appears lethargic, and her medical problems have gotten worse. Which of the following descriptions would apply to this home?

    a. Your mother was told upon entering, "Don't worry, sweetie, we'll take good care of you; we'll even tell you when it's time to go to bed."
    b. Your mother was told that the staff would take care of her physical needs, but she would be responsible for keeping her room arranged.
    c. Your mother was told that the nursing home would provide recreational activities; but if she wasn't satisfied, she would have to voice her concerns.
    d. Your mother was told upon entering the home that she could decorate her room in any way she saw fit.

Matching: Definitions

Match the letter of the term in the box with its definition.

a.	activity	i.	marriage
b.	Peck	j.	role exit
c.	disengagement	k.	constricted
d.	loneliness	l.	nursing homes
e.	integrated	m.	succorance seeking
f.	Erikson	n.	adult day care centers
g.	social exchange	o.	grandparents
h.	modernization		

1. psychologist who says that old age confronts men and women with three major tasks _____

2. term for elderly individuals with strong dependency needs and elicit responsiveness from others _____

3. psychologist who believes that the elderly confront the task of coming to terms with the issue of integrating versus despair _____

4. term used for elderly individuals who reveal a complex inner life, intact cognitive abilities, and competent egos _____

5. theory which views aging as a progressive process of physical, psychological, and social withdrawal from the wider world _____

6. often resemble volunteer firefighters in America; they are required to be on the scene when needed, but otherwise keep their assistance in reserve _____

7. theory stating people enter into social relationships because they derive rewards _____

8. term used for elderly individuals who structure their world to ward off what they regard as an imminent collapse of their rigid defenses _____

9. theory that views older people as having essentially the same psychological and social needs as middle-aged people _____

10. theory that regards the loss of occupational and marital statuses as particularly devastating, since these positions are anchoring points for adult identity _____

11. seems to play a role in protecting people from premature death _____

12. the awareness of an absence of meaningful integration with other individuals, a consciousness of being excluded from the system of opportunities and rewards in which other people participate _____

13. theory which assumes that the position of the aged in preindustrial, traditional societies is high because the aged tend to accumulate knowledge and control through their years of experience _____

14. the quality of life they afford is largely determined by bureaucratic requirements regarding Medicare and Medicaid and by state regulations mandating standards for licensing _____

15. one of the most promising alternatives to nursing homes; they offer part-time supervision for infirm and disabled individuals _____

Completion

Supply the term or terms needed to complete each of the following statements.

1. An elderly person who has come to terms with her own mortality experiences what Peck calls ego _____.

2. Neugarten would describe a _____-_____ as an elderly person who views aging as a threat and relentlessly clings to the patterns of middle age.

3. The best predictors of life _____ among the elderly are financial adequacy and good health.

4. According to the predictors of _____ theory, the elderly find themselves in a vulnerable situation because of the deterioration in their bargaining position.

5. When individuals are healthy and their incomes are adequate, they generally express satisfaction with _____.

6. Many elderly married couples say that companionship, respect, and the sharing of common interests _____ during later adulthood.

7. For many individuals, the grandparenting role is a source of biological _____ and emotional _____-_____.

8. Most psychologists, psychiatrists, and other health care professionals agree that, where possible, the elderly should be cared for in _____ settings.

9. In selecting an adequate _____ _____ for the elderly person, the following factors should be considered: the facility, safety, staff, activities, food, atmosphere, and cost.

10. Providing nursing home residents with a sense of _____ for themselves prevents some of the negative consequences of aging.

Multiple-Choice

Factual			Conceptual		
1.	b	(566)	1.	c	(566)
2.	a	(567-568)	2.	c	(566-567)
3.	d	(568)	3.	d	(567)
4.	b	(569)	4.	b	(568)
5.	b	(569)	5.	a	(570)
6.	c	(570)	6.	a	(570)
7.	a	(571)	7.	b	(571)
8.	c	(572)	8.	d	(575)
9.	c	(572)	9.	d	(575)
10.	c	(572)	10.	a	(578-579)
11.	b	(574-575)	11.	d	(579)
12.	a	(576)	12.	c	(581)
13.	d	(578)	13.	a	(583)
14.	d	(579)	14.	b	(583-584)
15.	a	(580)	15.	a	(585)
16.	c	(581)			
17.	b	(583)			
18.	c	(584)			
19.	b	(585)			
20.	d	(585)			

Matching: Definitions

1.	b	(566)	9.	a	(570)
2.	m	(567)	10.	j	(571)
3.	f	(566)	11.	i	(576)
4.	e	(567)	12.	d	(579)
5.	c	(569)	13.	h	(572)
6.	o	(580-581)	14.	l	(584)
7.	g	(571)	15.	n	(578)
8.	k	(567)			

Completion

1.	transcendence (567)	6.	improve (576)
2.	holder-on (567)	7.	renewal (continuity); self-fulfillment (581)
3.	satisfaction (568)	8.	home (579)
4.	social exchange (571)	9.	nursing home (582-583)
5.	retirement (575)	10.	responsibility (585)

Chapter 21

Death

INTRODUCTION

The concluding chapter of the text considers the antithesis of the life process--the stage of death. Several significant topics are discussed, including:

- The Dying Process. Modern definitions of death and how various societies perceive and recognize death are presented. How the elderly conceive of death, as well as the systematic psychological changes which precede death, is also examined. Kubler-Ross's stages of the death experience are outlined, and a critique of her stage approach is offered.

- The Search for "Healthy Dying." Research findings are presented regarding an individual's need to die with dignity. Salient features of the right-to-die movement are reviewed, and the controversial issue of euthanasia is discussed. The phenomenon of near-death experiences is explored. Within this context, the role of the hospice in comforting those near the end of life's journey is examined.

- Bereavement. Findings are revealed regarding typical adjustment to the death of a loved one. Finally, the coping experiences of widows and widowers are related.

I. The Dying Process

    A. Defining Death

    B. Confronting Death

    C. Dying

II. The Quest for "Healthy Dying"

    A. The Right-to-Die Movement

    B. Near-Death Experiences

    C. The Hospice Movement

III. Bereavement

    A. Adjusting to the Death of a Loved One

    B. Widows and Widowers

LEARNING OBJECTIVES

After completing Chapter 21, you should be able to:

1. Describe the changing American view of death. (page 590)

2. Summarize past and present definitions of death. (590)

3. Outline the ways in which elderly people confront death, specifically including the *life review* and death-drop. (590-592)

4. Discuss why *thanatologists* are concerned that people are ill-equipped to deal with the grieving process associated with death. (592-593)

5. List and briefly explain Kubler-Ross's stages of the death process. (593)

6. Explain Kastenbaum's reaction to Kubler-Ross's stage approach to the death experience. (593-594)

7. Define the following concepts: (594-599)

    a. healthy dying

    b. living wills

    c. the right-to-die movement

    d. *euthanasia*

8. Explain the relationship among stress, illness, and death. (Box, 596-597)

9. Discuss near-death experiences, and critique those descriptions. (599)

10. Explain the philosophy of the *hospice* movement. (599-602)

11. Identify the services provided to dying people by a *hospice*. (599-602)

12. Describe how the worldwide AIDS epidemic has created a plague mentality in some quarters. (Box, 600-601)

13. Define the following terms and discuss the functions and stages of development for each: (603)

   a. *bereavement*

   b. *grief*

   c. *mourning*

14. Identify and describe the four distinct stages of adult bereavement. (604-605)

15. Summarize the data regarding each of the following topics: (605-608)

   a. demographic statistics on widows and widowers

   b. the coping difficulties of widows and widowers

16. Discuss the importance of having a living will. (606-607)

271

KEY TERMS

bereavement (page 603)

euthanasia (595)

grief (603)

hospice (599)

life review (592)

mourning (603)

thanatology (590)

SELF-TESTS - CHAPTER 21

Multiple-Choice

Circle the letter of the response which *best* completes or answers each of the following statements and questions.

Factual Questions:

1. Thanatology involves

   a. the right to die
   b. clinical death
   c. life after death
   d. the study of death

2. Robert Butler indicates that the elderly tend to take stock of their lives and to reflect and reminisce, a process he terms the

   a. death-drop
   b. death-life film
   c. mirror of the life span
   d. life review

3. The death-drop phenomenon refers to

   a. the systematic psychological changes that occur before death
   b. intense guilt, despair, and depression that occur when one is told that death is near
   c. the otherworldly experiences reported by those who are resuscitated after clinical death
   d. the higher rates of death that follow significant life changes

4. Elisabeth Kubler-Ross distinguishes five stages through which dying people typically pass. In the middle phase, dying individuals try to arrange a truce with the illness in order to prolong their lives, and Kubler-Ross calls this the stage of

    a. anger
    b. acceptance
    c. depression
    d. bargaining

5. Dying people often begin to mourn their own deaths, the loss of all the people and things they have found meaningful, and the plans and dreams that will never be fulfilled. Kubler-Ross calls this

    a. thanatology
    b. denial
    c. preparatory grief
    d. life review

6. The death awareness movement asserts that

    a. life must be prolonged at all costs
    b. a basic human right is the power to control one's own dying process
    c. "aggressive" medical care should be used to prolong a dying person's life
    d. dying individuals never come to accept death

7. Another term for "mercy killing" is

    a. thanatology
    b. euthanasia
    c. senility
    d. convalescence

8. Richard Rahe asked men aboard three U.S. Navy cruisers to report their life changes and histories for the previous six months. During the next month, he found that the group of men who had been under the most stress had

    a. no illness
    b. significantly fewer illnesses than the group of men who had been under the least stress
    c. significantly more first illnesses than the group of men who had been under the least stress
    d. the same number of first illnesses as the group of men who had been under the least stress

9. An individual is most likely to become ill when he or she

    a. copes with stress
    b. is exposed to a perceived aggressor
    c. loses faith in religious healing
    d. doesn't experience enough life changes

10. Which of the following is likely to be reported by individuals with near-death experiences?

    a. passing through a tunnel and entering into an unearthly realm
    b. a decreased belief in God
    c. feeling as if the air is being squeezed out of their bodies
    d. intense feelings of blackness and desperation

11. Most hospice programs center upon

    a. attempts to prolong the life of the dying person
    b. developing cures for terminal illness
    c. "comfort care" rather than attempts to prolong life
    d. new experimental procedures designed to combat genetic disorders

12. Which of the following is the second leading cause of death in men aged 25 to 44?

    a. AIDS
    b. cancer
    c. heart disease
    d. suicide

13. Grief work involves

    a. mourning, talking about, and acknowledging the loss of a loved one
    b. heroic measures designed to prolong the life of a terminally ill patient
    c. placing a dying individual in a facility outside the home
    d. the reconstruction of new life patterns after the death of a loved one

14. Bereavement is *best* defined as the

    a. socially established manner of displaying signs of sorrow over a person's death
    b. state in which a person has been deprived of a relative or friend by death
    c. study of death
    d. individual's right to choose a self-actualized death

15. Helena Znaniecki Lopata studied a sample of widows in Chicago. She found that the widows whose lives were initially the most disorganized by their husbands' death were those

    a. from the higher socioeconomic classes
    b. living in ethnic communities
    c. from working-class families
    d. belonging to sex-segregated worlds

Conceptual Questions:

1. Absence of spontaneous breath is to _____ as spontaneous brain function is to life.

    a. thanatology
    b. death
    c. life
    d. artificial ventilation

2. It seems that lately all your elderly mother wants to talk about is her childhood in Poland and her immigration to the United States. Your *best* interpretation of your mother's behavior would be that she is

    a. morbidly depressed and preoccupied
    b. regressing to a more primitive developmental level
    c. preparing herself for a lonely, debilitating death
    d. exhibiting a normal and healthy form of personality development

274

3. Your grandfather is a lovable old man who will stand his ground and argue just for the sake of arguing. However, over the last few months, he seems to have become more apathetic and introverted. We might reasonably predict that

   a. his mental abilities have declined as well
   b. he has just recuperated from a serious illness
   c. he is in the advanced stages of Alzheimer's disease
   d. his growing introversion is a normative change in the elderly

4. An elderly man is informed that he is suffering from a terminal illness. The man tells his doctor, "You must have mixed up my lab results with someone else's." This response would be characterized by Kubler-Ross as

   a. bargaining
   b. denial
   c. acceptance
   d. rejection

5. Intolerance of pain is to _____ as a living will is to euthanasia.

   a. the right to die
   b. healthy dying
   c. depression
   d. living wills

6. A man who works in your office just had a major heart attack. Everyone who knows the man is amazed because he was in perfect health and had no history of heart problems. A reasonable conclusion might be that the man

   a. was depressed because of a recent negative life review
   b. underwent a major personality change in the last year
   c. was trying too hard to get a promotion
   d. experienced an unusual number of life changes recently

7. Your elderly aunt tells you, "If I ever have a stroke, I don't want you to keep me alive with tubes and needles and machines." This statement

   a. would by unconditionally accepted by doctors
   b. would be approved of by most religious groups
   c. is a death wish that the elderly make when they resign themselves to dying
   d. might not be accepted at face value by a doctor

8. Your car plunges off an icy bridge, and you are trapped under the water. Although pronounced dead on arrival at the hospital, you are miraculously resuscitated. Later, all you can recall of this incident is plunging into the water. Your reaction

   a. was an out-of-body experience
   b. offers evidence for a spiritual existence beyond death
   c. does not support the notion of a life-after-death phenomenon
   d. was probably produced by a hyperventilated brain

9. You have talked to your elderly mother's physician about her failing health. Although he is concerned about her comfort and well-being, he seems more concerned about being a medical failure and losing the battle to cure her illness. The doctor's attitude is inconsistent with which of the following concepts?

   a. medical model
   b. hospice
   c. bereavement
   d. right to life

10. The best slogan to assist young, sexually active people in dealing with the AIDS epidemic would be which of the following?

   a. Don't think twice; it's all right.
   b. Pay me now, or pay me later.
   c. Nothing ventured is nothing gained.
   d. If you want to dance, you have to pay the band.

11. Your son's little friend was hit and killed by a car. You are unsure about whether your son should attend the funeral. A thanatologist might tell you that

   a. attending the funeral will give your son death-related anxieties
   b. if your son does not attend the funeral, he will soon feel that his friend has departed
   c. attending the funeral will help your son see death as a natural end to life and form beliefs about it
   d. attending the funeral will just raise more questions in your son's mind than it answers

12. Your neighbor's husband died recently. Despite her loss, she has shown a positive adjustment to her new circumstances. She is also beginning to leave her home more often to get reacquainted with people and activities. She is currently in which phase of the adult bereavement process?

   a. first phase
   b. second phase
   c. third phase
   d. fourth phase

13. Your father died four years ago. Your mother still has all your father's belongings and has left them untouched. She has turned into a depressed recluse, relying more and more on you for support. Psychologists would say that your mother

   a. had not progressed beyond the earliest stages of the grieving process
   b. would have resolved her grief more effectively if your father had died suddenly
   c. had begun the identity reconstruction phase of the grieving process
   d. had passed through the emancipation stage of the grieving process

14. Which of the following vignettes would we be the *most* likely to observe?

   a. An 80-year-old woman is still living with her husband of fifty years.
   b. A 65-year old widower remarries two years after the death of his wife.
   c. A widower develops new male acquaintances after the loss of his wife so that he can share his loneliness with others.
   d. A 65-year old widow decides to move in with her married daughter's family.

15. Mary is a 75-year-old, recently widowed woman who was a devoted wife. She has a college degree, and her late husband was an attorney. Which of the following predictions could most reasonably be made about Mary?

    a. She is more likely to remarry than a same-age widower.
    b. Mary probably will prefer to maintain her kinship links by moving in with her married daughter.
    c. She will be at high risk for suicide because of unresolved feelings of loneliness.
    d. Mary will idealize her late husband.

Matching: Definitions

Match the letter of the term in the box with its definition.

a.	stress	i.	life review
b.	comfort-care	j.	hospice
c.	thanatology	k.	death-drop
d.	grief	l.	bereavement
e.	education	m.	living wills
f.	denial	n.	AIDS
g.	mourning	o.	euthanasia
h.	preparatory grief	p.	near-death experiences

1. socially established manner of displaying signs of sorrow over a person's death _____

2. increasingly has been implicated as a causal factor in illness and death _____

3. the nation's lone "vaccine" against AIDS _____

4. the study of death _____

5. systematic psychological changes which occur before death _____

6. second leading cause of death in men aged 25 to 44 _____

7. mercy killing; putting to death painlessly an individual suffering from an incurable and painful disease _____

8. a process in which the elderly take stock of their lives, reflecting and reminiscing about themselves _____

9. a stage the dying person goes through in which the individual resists acknowledging the reality of impending death _____

10. these are commonly precipitated by a medical illness, a traumatic accident, a surgical operation, childbirth, or drug ingestion _____

11. dying people experience this when they begin to mourn their own deaths, the loss of all the people and things they have found meaningful, and the plans and dreams that will never be fulfilled _____

12. legal documents that give an individual the right to refuse "heroic measures" to prolong his or her life in the event of terminal illness _____

13. a state in which a person has been deprived of a relative or friend by death _____

14. keen mental anguish and sorrow over the death of a loved one _____

15. involves an aggressive treatment of symptoms, both physical and emotional, through the use of counseling, antidepressive medications, and high-dose morphine _____

16. A program or mode of care that seeks to make dying less painful and emotionally traumatic for both patient and loved one _____

Completion

Supply the term or terms needed to complete each of the following statements.

1. The earliest Biblical sources considered a person's ability to _____ independently to be the prime index of life.

2. Kastenbaum's criticism of Kubler-Ross's theory revolves around his contention that she ignores the _____ of the disease itself in regard to the death process.

3. The purely _____ model of disease based on the notion of "one germ, one disease," and one therapy is recognized as much too simplistic.

4. The _____-_____-_____ movement claims scientific evidence of a spiritual existence beyond death.

5. _____ does not spread through casual contact.

6. Advocates of the _____ approach say that it is difficult for physicians and nurses taking care of patients in hospital settings to accept the inevitability of death.

7. Unfortunately, cultural expectations, social values, and community practices at times interfere with necessary _____ _____.

8. The _____ phase of adult bereavement involves pining, yearning, and depression.

9. The Cruzan decision, emphasizing the need for "clear and convincing" proof of a patient's wishes, has encouraged greater use of _____ _____.

10. Research has suggested that the negative long-term consequences of widowhood seem to derive from _____ deprivation rather than from widowhood itself.

Multiple-Choice

Factual

1.	d	(590)
2.	d	(592)
3.	a	(592)
4.	d	(593)
5.	c	(593)
6.	b	(594)
7.	b	(595)
8.	c	(596)
9.	a	(596)
10.	a	(599)
11.	c	(599-560)
12.	a	(600)
13.	a	(603)
14.	b	(603)
15.	a	(608)

Conceptual

1.	b	(590)
2.	d	(592)
3.	a	(592)
4.	b	(593)
5.	a	(594-595)
6.	d	(596)
7.	d	(598)
8.	c	(599)
9.	b	(599-602)
10.	d	(600-601)
11.	c	(603)
12.	c	(604)
13.	a	(604-605)
14.	b	(605-608)
15.	d	(608)

Matching: Definitions

1.	g	(603)
2.	a	(596)
3.	e	(601)
4.	c	(590)
5.	k	(592)
6.	n	(600)
7.	o	(595)
8.	i	(592)

9.	f	(593)
10.	p	(599)
11.	h	(593)
12.	m	(594)
13.	l	(603)
14.	d	(603)
15.	b	(600)
16.	j	(599)

Completion

1.	breathe	(590)
2.	nature	(593)
3.	medical/biological	(597)
4.	life-after-life	(599)
5.	AIDS	(600)

6.	hospice	(602)
7.	grief work	(603)
8.	second	(604)
9.	living wills	(606)
10.	socioeconomic	(608)